*"When you care enough about
Something, you speak about it.
And do Something.
Action starts with speaking"*
.......***Anonymous***

"One who doesn't care, doesn't count"
........***Anonymous***

"The 'Beautyful' ones are not yet born"
......***Ayi Kwei Armah***

*"Disregard for the past will never do us any good. Without it
we cannot know truly who we are."*
.......***Syd Moore***

DUE SEASON

TO WHOM IT MAY CONCERN

Fola Soboyejo

For additional copies of this or other Fola Soboyejo titles write:
Korloki Publishing Company Inc.
1320 Coney Island, Unit B3, Brooklyn, NY 11230.
Please allow 4 to 6 weeks for delivery.
For bulk orders contact us via email kpcbooks@yahoo.com

Cover design by: Lisa Bracken
Interior design: Allzents Groups Inc.
Summary: A Memoir

ISBN13: 978-1-936-73932-5
ISBN10: 1-936-73932-1

Printed in the United States of America

DEDICATION

Dedicated

to

mom Omolara and late dad Adekunle

for nurturing with love and discipline.

And to all who care about humanity and development in Africa

for the struggle continues; the best is just around the corner.

TABLE OF CONTENTS

THANK YOU

This is to thank all friends, colleagues, and family for the inspiration, suggestions and support. My special gratitude goes to my wife Titilope for producing and reproducing the manuscripts and indulgence of the odd hours this work demanded. Without Korloki Books publisher – who keeps telling me – 'it is easier than you think' - **Due Season** would just be in my head for much longer. Kudos.

DUE
SEASON

BY

FOLA SOBOYEJO

FOREWORD

This remarkable Collage of Essays and Presentations in Nigeria and Canada 'Due Season' is a single bright sword that cuts across the diverse subjects of International Partnerships, Engineering, Economics, Business decision-making and Ethics which can bring insight and reading enjoyment to researchers, students at all levels, and broaden the perspectives of professionals, entrepreneurs, policy makers and lay persons.

International Partnerships promote trade and bring about inward investment to tap opportunities in growing economies like Africa. Chapter 1 discusses the need for partnerships, their benefits, and safeguards to overcome pitfalls, and ensure success.

General insecurity of lives and property in Nigeria — and many other resource — rich countries - is of great concern to government and the governed. The collision of mineral exploitation corporations and restive local communities has been on the increase. Chapter 2 gives particular insight into the issues of security and safety in the Petroleum and Allied Industries Sector and proffers solutions to the lingering and imminent problems. The issues addressed and postulations made by the author many years ago are still valid today.

Operators of Small and Medium Enterprises in Nigeria have common challenges in prudent decision-making and risk-taking which hinder business growth. In order to tackle these challenges, Chapter 3 offers tested management accounting tools and a simple illustration that can be used for production and profit planning by SMEs in the manufacturing or service sectors.

The Petroleum Sector is the major foreign exchange earner for Nigeria. The continued marginalisation of other sectors by Government policy and inaction makes the country's economy vulnerable to the vagaries of the volatility of international demand for oil, and increasing discovering of oil reserves in different parts of the world in general and huge shale oil development in the USA. Chapter 4 discusses strategies for Nigeria to stabilize the economy by lifting her non-oil foreign exchange revenue contribution from the present infinitesimal level to 50% of total export.

The world is a global village and there is need for networking amongst professionals.

Chapter 5 emphasizes this fact and its benefits particularly between Canadians and

Nigerians to tap into the synergy between high technology and expertise with local market skill and experience.

Moral values are on the decline these days the world over even in faith movements. On the other hand, trust and credibility are in constant demand for meaningful social and commercial life of individuals and businesses. Chapter 6 underscores the essence of integrity and provides guidelines and tools that help in 'doing the right thing'.

This Collage could not have come at a more auspicious time than now when the author is marking his Diamond Anniversary on earth. He has great passion for learning and imparting ideas for a better, productive and equitable society with his God-given talents. To go beyond the realm of ideas, he is moving forward with 'love in action' which is the motto of the Charity Foundation he has just founded - to give hope and assistance in alleviating educational and rural community decline in Africa.

I have been privileged to collaborate with the author in professional societies, have attended some of his presentations, came to the conclusion that he has complete grasp of all the topics and issues. As the issues and ideas are true for Nigeria, they are true for other countries.

Without reservation, I commend his candor and succinctness.

It is remarkable that he is committing proceeds from this book to causes of the SOBOYEJO-TURNINGPOINT Charity Foundation, which I commend to everyone to support and help find turning points for rural communities in educational and socio-economic decline in Africa.

Due Season is a book to have, and a must read for all and sundry.

—Engr. Olusegun Oyelola, FNSE
Lagos Nigeria

TO WHOM IT MAY CONCERN - is a sub-title for **Due Season** as the first book in a series by the author on a variety of issues and visions on development, the human experience, the immigrant psyche and stories of mixed bag of economic well-being, broken dreams, loss of identity, and even hope swings. Whereas the series may be predicated on the African perspective, **To Whom It May Concern** opens the window of light to all who are, or are involved with newcomers in different parts of the world.

DUE SEASON – is a Collage of Essays and Presentations in Nigeria and Canada by Fola Soboyejo on a variety of topical issues of Development and Ethics, published on the occasion of 60 PRAISES - for Fola's 60 years in the world of the living. To bless back, being so blessed himself, proceeds from **Due Season** go to Ogan Village School (Broken Classrooms Re-development) Charity Project in Ifo Local Government Area, Ogun State Nigeria and subsequently to SOBOYEJO-TURNINGPOINT CHARITY FOUNDATION for Educational and Rural Development in Africa – giving hope and help to alleviate educational and rural community decline.

Contents

DUE
SEASON

BY

FOLA SOBOYEJO

INTRODUCTION

From college days experience in the Press Club, to my time in Unipetrol / Esso when I enjoyed the support of my boss to attend conferences, etc., article and paper writing on things I cared about, was always nerve-racking and time consuming. But I enjoyed it.

Each time I picked my past papers, considering the research that went into the papers, and articles, not to talk of the zeal, burning desire and energy that went into them, plus many of the concepts, conclusions, strategies, and recommendations which years after, are still valid today, I found myself wanting to pass on something to fellow thinkers, compatriots and future generations.

The details of the issues and solutions have changed but the trend is the same. Overall development around the world and particularly Africa over the last five years reinforces the promise that transformation has begun and will come to full bloom and boom in time. When it is time. Hence the book title **DUE SEASON**. When things change and trend remains, if we can manage the high and low tides, we will emerge.

If the argument 20 years ago and today is still about national commitment, vigilance, sacrifice, home-made goods and services, collaboration of home professionals and those in the Diaspora and self-reliance for emancipation, so be it. Posterity, techno-economic analysts and players of different generations may learn something from the trends enunciated in this book, and not allow history to repeat itself. Committed policy makers and implementers may actually use their knowledge of trends and history to break the vicious cycle of underdevelopment.

May this work, the chronicles of the ideas in the articles, papers and presentations, as a symbol of service, and vision, be an inspiration to all who read it. Records tell a story that can throw up red flags or green lights, that integrity and commitment are possible, that if we can navigate the pitfalls and rise above the vicious cycle (in case we cannot break it), then progress and prosperity for our people are possible.

After reading each chapter, I believe you will see that you have gotten something out of this book. Much more importantly, you will come to realization that we can emerge, Nigeria can emerge, and Africa can emerge and lead the world again. Yes, We Can (in the words of Barack Obama on the US Presidential campaign 2008). Yes, indeed we will in due season, if we don't relent.

Chapter One

INTERNATIONAL PARTNERSHIPS – NOT FOR THE COWARD OR THE NAÏVE

PURPOSE

To inform and kindle the mind of the audience on the essence, barriers in the way of international partnerships for direct inward investment in opportunities brought about by Africa's high economic growth rate, and show how to navigate from dream to reality.

PREAMBLE

African economic growth rate is one of the highest in the world for the first time in history. This means opportunity for high-yield investors from around the world to invest or have business partnerships where there is growth.

From the World Bank to OECD (Organisation for Economic Cooperation and

Development) and the African Development Bank, there is a concert of statistics that confirm that the emerging economies of Africa recorded the highest GDP (Gross Domestic Product) growth rates in the last five years. The high growth rates were the highest ever not just for the continent, but superior to other emerging economies in Asia and Latin America. The outstanding economic growths also superseded the growth rate of Europe, the United States of America and Canada.

Year 2011
Average GDP growth rate for **Africa – 7.5%**
Asia, except China – 5%
Latin America – 4%,   USA – 1.5%,   Canada - 2.5%
Ghana – 14% Nigeria – 8.4% China – 11%

The impressive African economic performance can be attributed to commodities, mining, agriculture and the resurgence of the middle class. This new trend of improvements is not unlike the development that was experienced in other parts of the world where growth has been translated into better life for the people and prosperity for the countries. In recent memory, growth in the Asian

Tigers countries significantly pushed up per capita income and the standard of living.

Why is the growth in GDP of African countries not adequate to positively impact the people – from Soweto Township (Johannesburg) to Ajegunle (Jungle City in Lagos) and shanty town in Nairobi? The simple answer is that real productive investment especially direct inward investment is an important part of what is required for a populous continent with vast natural resources to significantly improve the lives of the people..

In an intricately interconnected world, the recipes for prosperity are many. They include production, sizable consumer market, balance of trade, and direct inward investment.

FOCUS

Our focus here is to x-ray one of the factors for prosperity - direct inward investment, as key to modern technological development - through partnerships between foreign investors and local entrepreneurs and professionals. The intrinsic issues around ability, capacity, motivation, pre-conceived notions, and handicaps of the local and foreign partnering parties, as well as the extrinsic barriers beyond the control of the parties will be discussed. Solutions to overcome the issues in the way of international partnerships will be proffered to transform desire to reality of economic prosperity for the people of Africa, with unbeatable returns to its international foreign investors.

Foreign investment opportunity in Africa with a market of 1 billion people is worth trillions of dollars over several years for the investors and their local partners with potential for higher standard of living for the people. But why is this potential almost like pipedream?

There are barriers against direct inward investment contributions to development in Africa that must be surmounted.

WHAT ARE THE BARRIERS

There are numerous barriers to direct inward investment in Africa. The barriers can be described in broad terms as perceived barriers, and real barriers.

Perceived barriers can be described as myths and stereotypes. These barriers are not totally imaginary issues. In fact they do have elements of truth or history. They are often exaggerated. Let's look at a few examples.

PRE-CONCEIVED NOTION / STEREOTYPE 1

'Africa is backward – ridden with disease, hunger, ignorance, and the people are rural'.

To say Africa is none of these is wrong, and to say all Africa is all of these is also wrong.

The truth can be found in the records of the UNICEF, UNDP, WHO, FAO, and other agencies that work with the countries and know how much Africa has changed over the last two decades. Infant mortality has reduced by half, literacy increased by half, etc.

But when the perception in western press is all about an outbreak of drought, flooding, cholera, kwashiorkor, and corruption, who will be excited to go to such a place? Breaking news is bad news. Breaking new is big news, big news is big money.

Prime time advertising means revenues and profits. It's all about money. For a section of the media, Africa is not on their radar. Hearsay and made up stories is their stock in trade

As a result Africa must speak for itself. African embassies and the African Union must blow their own trumpets, and direct attention to continental change, growth, expanding pockets of development and tremendous opportunities for substantially more direct inward investments with billions of dollars to be made.

The Continental Africa Chamber of Commerce USA in Chicago Illinois is making a wonderful trade promotion contribution in this regard

PRE-CONCEIVED / STEREOTYPE 2

'In Africa, there is no security, but war all over the place, compounded by piracy and terror.'

Nobody would like to leave a place where there is reliable policing, rule of law, public safety and security and take their investments to a place that is lacking in all these factors of human comfort and peace of mind. Of course, there is no defence for piracy on the coasts of Africa – Indian Ocean and recently Atlantic Ocean. There is no defence for continued war in the Congo basin and genocide in Darfur. It is untenable to have unabated cases of bomb blasts in Nigeria and Kenya which had enjoyed tranquility for decades. Thank God there is a semblance of peace since Southern Sudan won its Independence, and ECOWAS (Economic Community of West African States) is helping to stabilize Mali, just as they succeeded along with the French army and the UN team in Cote d'Ivoire to attain peace a few years ago

On a global basis, the question is, Where in the world is safe?

The US was safe until the Twin Towers were brought down on 9-11-2001, in a gruesome plane attack that took over 3,000 lives. That was an unprecedented breach of peace and security in human history. Britain was safe until coordinated attacks on the London Underground trains that killed hundreds of people on July 7, 2005. Canada was safe until a gunman – Marc Lepin killed 28 College students in the Montreal massacre 1989.

Street gang war in the US and Canada has now graduated to cannibalism where human body parts were found floating in city rivers in recent years. More gory still, an adolescent son of a professor killed and plucked and ate the heart of a Ghanaian doctorate student living with his family in Maryland USA. Police brutality and stray bullet investigations are prevalent. These sound like places that are too scary to live in or establish business, but..........

Therefore, security is all relative; in time and places.

SECURITY AND DEVELOPMENT

Next is the issue of security, investment and development. Let us consider the comparative level of violence in Latin America and direct foreign investments versus the situation in Africa. It is a pity that proper reporting on crimes and accidents in Africa is lacking. But if we exclude the war front countries, cases of loss of human lives from homicides and open violence are extremely low in Africa.

By comparison with most of South America, Africa appears to fair well. For instance, it is well known that Honduras has the dubious reputation of being the murder capital of the world, while over 10,000 people are killed annually in Mexico's drug wars. Yet, Latin America still enjoys huge foreign investments relative to Africa.

TABLE 1: CANADIAN / US TRADE WITH AFRICAN AND MEXICO			
	2009	2010	2011
US - MEXICO	$304Billion	$393B	$450B
US - AFRICA	$87B	$113B	$127B
CANADA- AFRICA			$8B

As a result of this trend, one can ask: Is there justification for pre-conceived notions, mischaracterization, stereotyping, isolation or neglect of Africa when it comes to investments from western countries? Comparatively, Africa is **not** more violent, deserves more attention and more investments than is presently the case. The market is large. The middle class is large and growing. Disposable income is increasing. There is huge infrastructure development deficit, which means a huge investment scope. African governments, the press, embassies, the Chambers of Commerce and the African Union and their foreign allies must continually campaign that Africa is a good place for foreign investors to come, establish, and prosper.

PRE-CONCEIVED NOTION AND STEREOTYPE 3

'Corruption is huge in Africa' appears to be the swan song in the West.

But the quick question is, does it not take two to tango?

Corruption is a cancer that destroys the ethical fabrics of commerce and society and escalates the cost of doing business. It creates an unattractive environment for foreign investment and international partnerships in general.

Again there is no justification or defence for bad governance, weak regulations and law enforcement, treasury looting, siphoning of resources, graft, and stashing peoples' funds away in Paris, London, Zurich, New York or Toronto; even Hong Kong or St. Petersburg. Shame is shame. Shame unto those corrupt African leaders, legislators and business leaders. The irresponsible behaviour of such leaders is a huge setback for African development and especially trust – which is the sine qua non for meaningful partnerships.

However, what is the role of developed countries in this global heinous sport of corruption, and the set back of Africa?

It is an open secret that Switzerland is the international vault of ill-gotten gains. Swiss banks with their secret coded accounts are gold medalists among custodians of stolen money. Only two years ago the administration of President Obama picked battle with UBS - Swiss bank and forced disclosure of the off-shore accounts of 20,000 tax dodging American billionaires who lodged over $20B in the bank's vault.

The same secret account scheme for looters and tax dodgers – in Cayman Island, Channel Islands, Monaco, etc., is officially propped by western governments who turn blind eyes to those safe havens. So, western countries just lead the way and their African cohorts follow suit in financial crime cover-ups

On yet another note, it is barely a year ago that the head of Nigeria's anti-graft agency—Mrs. Farida Waziri indicted former American Vice–President Dick Cheney on serious charges of corrupting Nigerian public officials and politicians in the Halliburton-Kellogg bribery scandal. The V-P settled out of court and refunded millions of dollars to Nigeria. On the one hand, it is curious and inexcusable that the V-P's Nigerian comrades-in-crime were not tried nor jailed. On the other hand, the question is what happened to American justice under the Corruption of Foreign Officials Act? Is the V-P a sacred cow?

According to an African folks-saying, 'It is not the one who snitched from the attic that is the thief, but the receiver of stolen goods'. So between the looters of African treasuries and the receiver banks and countries, who is the thief. Who is corrupt? Who is clean?

The point to note is that it takes two to tango. So, corruption must be tackled by both the treasury looters' home countries and the loot receiving and banking countries. The holier than thou attitude in trade and investment dealings must stop. African political leaders, business leaders and institutions must put their house in order to earn respect and good international credit ratings. Ethical conduct and fair treatment must prevail for meaningful international partnerships and abundant direct inward investments.

With the myths and stereotypes put in proper perspectives, hopefully tackled head on, the way is paved for level playing field for African countries among other developing regions vying for foreign investment. It may also create a refreshing opportunity for respectable genuine trade and investment relations with Africa.

But what are the prerequisites to bridge the billions of dollars available for investment into developing countries, particularly Africa. The reputable journal – (London) Guardian Weekly – in the 31 August 2012 edition, said so much about this subject in a detailed article titled, 'Our (western) image of Africa is hopelessly out of date'.   We believe that African media and voices should echo and sound the loudest in matters of such affirmative declaration.

10 CRITICAL STEPS

From Dream to Reality - Securing International Partnerships

The Scripture admonishes, 'No one set's out to build a house without first planning (counting the quantities and costs)', if a project or business is to be successful.

In a particular small business consulting service, hardly does a month pass without receiving business inquiries from Africa for trade, investment, and technical partnerships. The inquiries are usually high-sounding, and fascinating. But so many errors and omissions make many trade requests unsuccessful. Our analysis over the years led to the following criteria that must be satisfied to have a chance for investment partnership opportunities.

1. Preparation is everything – why, when, where, what international partnership?
2. Draw Business Plan – including SWOT (strength, weakness, opportunity , threat) Analysis and PEST (political, economic, sociological, technological) Analysis; Scripture asks, 'Is there anyone planning to build a house who would not figure out what it takes?' **Luke 14:28**
3. Business Incorporation–calls for more rigorous accountability and transparency to the law and to lenders and investors;
4. Research!!! – the market, investment incentives, preferential sectors, appropriate technologies, regulations and standards – local and international
5. Strong Credit Profile – local, then international profiles – ensure credit line qualification
6. Pre-Feasibility Study including Risk Assessment – what if? Scenario building
7. Due Diligence (especially by the foreign partner)– engage local / Diaspora/ international consultants for intelligence - local people, environment, the line between gifting,

lobbying versus bribery, cultural / social etiquette, values, and the human face

8. Pre-Visit the investment destination: discoveries may surprise you; gives opportunity to engage stakeholders, validate claims, land/asset title, pre-qualifications; be engaged – connect with chambers of commerce, trade groups, professional associations

9. Evidence of readiness – practically assess, prove / disprove land / assets, capital contribution commitments, project RFPs (request for proposal) as may be applicable.

10. Pilot scale project before full scale collaborative project – partnership in action, on trial basis.

CHINA

A model in International Partnership and direct inward investment.

In the same circumstances of Africa's shortcomings and constraints, when compared with western countries the Peoples' Republic of China is now the global leader in direct inward investment in Africa. Of course China is entitled to its own political and imperial ambitions and mineral supplies security objectives. The point is, China's investment in Africa has overtaken that of the traditional colonial and post-colonial allies, in mineral development, railways, roads, and other infrastructure sectors.

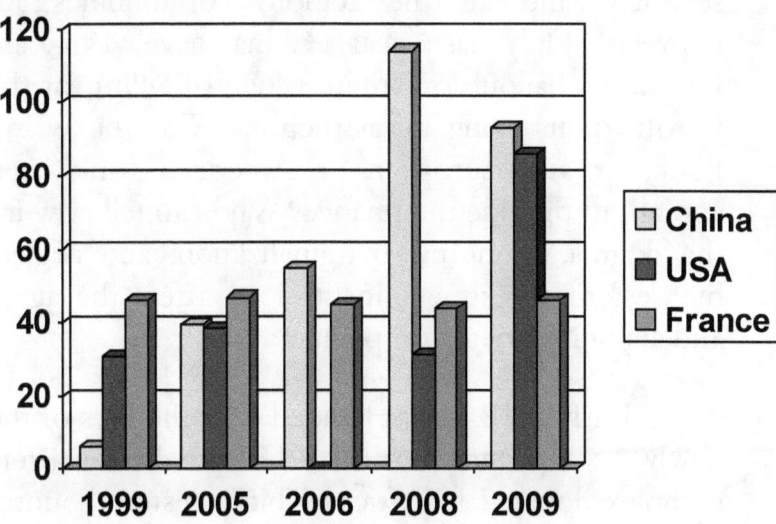

Fig.1: Africa Direct Inward Investments ($billion)

Talking about Chinese support when African projects are written off as being unviable by the West, way back in the 60's and 70's China financed the Tan-Zam 1,860 Km railways – the first *international* rail link on the continent - between landlocked Zambian Copperfields and Tanzanian seaports, and all the remote communities in-between. Only last year, China gave Tan-Zam Railways a bailout low interest loan of $49m for the retrofit of its aging infrastructure. Was the World Bank, or any European or American bank keen enough to provide the finance? Who can tell how far the dogma, economic marginalization, and neglect by western investment institutions affect the aging and ailing Nigerian steel plants?

Early 2012, China handed over the keys of the newly completed African Union headquarters complex donated to Africa. Which western country would do that in spite of centuries of relationship? What was China thinking? What are western countries thinking? Time will tell. . .

STRATEGIC DEVELOPMENT AND PARTNERSHIP PLATFORMS

About 70% of the remarkable growth rates in Africa comes from commodities – raw minerals and produce – gold, oil & gas, cocoa, banana, platinum, uranium, copper, etc. If the natural resources continue to be carted away, the local population will be deprived of real development. If a substantial part of the natural resources is processed and converted to finished products, there will be gainful employment, higher incomes, and a bigger richer populace to constitute a stronger consumer market for a variety of modern and technological goods. Again there must be a roadmap to achieve this. We therefore propose that the African Union must design, proclaim and implement a plan that we call the **African Strategic Development Agenda** based on 6 platforms:

1. Self-Reliance – Food, Energy – Power, Oil & Gas, Renewables, and resource processing
2. Economic infrastructure: Power plants, Refineries, Potable water, Irrigation, Roads, Railways, Communication, the Internet

3. Human development services – Education, training, health, sports and culture
4. Centres of excellence for Research & Development – Technology, Arts and Sociology
5. Human Rights, labour rights, 1 man 1 vote, maximum 8 years in political office
6. Afrocentric partnerships

EXTRINSIC BUSINESS BARRIERS AND SOLUTIONS (By Governments)

1. Rule of law and Security of lives and property
2. Political stability – constitutional rule, consistent 2 - term policy, devolution of power from the centre to strengthen the geo-political zones or regions
3. One-Stop Shop for business information, registration, incentives, regulations, local content policy, Made In Africa/ Nigeria / export promotions
4. Good governance – accountability and transparency in tenders, etc.
5. Economic reforms – genuine deregulation, e.g. privatisation of the national grid

6. Huge investment in health and education, especially technical, enterprise management, and business tax rebate for youth employment
7. Social justice - Free press, human rights and civil society institutions, living wage
8. Aggressive rural economic development programmes, including roads, potable water and cottage industries
9. Regional economic integration – transportation, ECOWAS free trade zone with the removal of customs barriers, single currency – introduction of the 'Afro' in 2023 through AfDB (African Development Bank) and proposed African Central Bank.
10. Justice: Jail economic criminals, recover the loot from Switzerland, Paris, New York, London and other places; encourage new advocacy by Transparency International and allies - that will enable whistle-blowers to upload Online the occurrence of corruption, the looters, and the loot receiving institutions and countries.

ATTRACTIONS AND OPPORTUNITIES
- for International Partnerships with Africa

1. Africa's Population – 1 billion (census and estimates to 2010)
2. 60% population under age 25, 7-10m/year youth entering labor market
3. 50% literate society; democratic countries 49 out of 54
4. Large /growing middle / working class – about 500m people of working age – untapped potentially productive human capital resource and solid purchasing power
5. Great number of African accomplished professionals in the Diaspora, ready to plough back, serve tenures, and stay in Africa
6. Lower cost of labor - even at professional level a Project Manager in Africa will cost you $35,000 while counterpart from the same class and program will cost you in North America $60,000 (PMI); Wages rising faster in China than in Africa
7. Lower cost of living, higher perks value for executives and expatriates

8. Lower population density than many developed and developing regions of the world
9. Market economy, growing private capital Stock Exchange, more traded companies, stronger SMEs, investment banking, community banks, insurance.
10. Huge reserves of resources – gold, coal, uranium, oil & gas, iron & steel, titanium, etc
11. Business enabling environment – deregulation, privatisation, one-stop shop for new business establishment, expatriation of profits, computerised customs and immigration processes, tax holidays, export incentives
12. Private-Public-Partnership (PPP) projects development strategy – policy in practice
13. Infrastructure – $93b infrastructure development deficit a year (ref. World Bank), for the next 20 years - e.g. highways, railways, aerodromes, water, power, solar energy;
14. Nigeria investing $7b by 2014 on infrastructure, Libya investing $500b by 2021

VIABLE SECTORS

Top of the viable investment sectors are: Health, Education, Food, Agro-Allied, Cloud Control Artificial Rain-making (applied in China and Middle East deserts) to reduce desert encroachment, Housing, Tourism, Music & Movies, Consumer Electronics, Personal Products, Automobiles, Oil & Gas, Communication & IT, Engineering and Project Management Consulting, Green Energy, Environment management and remediation, Iron & Steel, Bauxite, Precious stones and metals iron ore, tin, gold, silver, bauxite, copper, uranium, molybdenum, diamonds, (clean) coal;

Manufacturing – huge scope - as % of GDP remained unchanged between 1995 – 2004 – 6% compared to 11% in other developing regions of the world (AfDB) – a great scope of opportunity for investment.

PRODUCTIVE CAPACITY

With the wave of privatisation and government pull back from direct investment and increasing deregulated environment, the only way real development can occur; prosperity created with positive impacts on the people is through international partnerships for foreign investment. In order to create productive capacity on the continent for original equipment manufacturer, process plant design and construction for level of technology- and cost-intensive facilities such as petroleum refineries, mining process plants, power generation, and communication / IT software and hardware, foreign investment is the key.

CAN - DO ATTITUDE, CULTURAL PRIDE AND DETERMINATION

According to a saying, 'If you are not sure, in order to know where you're going, look back where you are coming from, stand steady where you are, with the roadmap, and with confidence the only direction before you is Forward".

Despite the chequered African experience from slavery to colonialism, imperial exploitations, disqualifications and marginalisation, African dignity, culture and pride is unquenchable. In the quest for rapid development and the search for partners on the challenging journey ahead, it is motivational to mention some of the past and current African innovations and inspirational achievers.

Innovations – hieroglyphics, mathematics, metallurgy, pyramids, sphinx, astronomy, engineering, chemistry, medicine, mummies, warfare armament, and political administration.

Inspirational figures - Ibn Battuta – Moroccan scholar and traveler – 14th century (visited China 1345), Chaka - The Zulu, Mansa Musa, Usman Dan Fodio , Oduduwa,– with descendants in West Africa and Brazil, Dr. Julius Nyerere – Tanzania's illustrious President, people's leader Dr. Kwame Nkrumah, John Garanje – South Sudan leader whose efforts paid off after his death, Nobel Laureates – President Nelson Mandela (South Africa), Dr. Wangari Maathai (Kenya), Prof Wole Soyinka (Nigeria). UN Sec.–Gen. Kofi Anan (Ghana).

Other African inspirational figures are Chief Obafemi Awolowo, Playwrights Chinua Achebe (Nigeria), Zakes Mda (South Africa), Economists Dr. Ngozi Okonjo-Iweala (Nigeria), Dr. Dambisa

Moyo (Zambia) and a number of women presidents, ministers, professors and entrepreneurs.

Fig. 2: New 20-Storey African Union Hall, Addis Ababa – US$200m donated by China on the 10th Anniversary of the union, January 2012

Quote:
"At least the Chinese are willing to buy your kisses, unlike Western countries who demand your kiss with guns and bullets, and left you with poverty and indignity" – John Bablar, Addis Ababa Online (on evolving eras of exploitation - from slave trade through colonialism to neo-colonialism, and custody of stashed ill-gotten wealth of corrupt leaders and their accomplices).

CONCLUSION

1. In an intricately interconnected world, modern development depends on direct inward investments made possible through international partnerships.
   Africa's 1 billion people with increasing educated youth and expanding middle class with growing purchasing power offer huge market of trillions of dollars and attractive returns for local and foreign investors.

2. For local professionals and entrepreneurs to attract international partnerships they must be prepared with homework, organization and transparency.

3. Foreign businessmen who want to tap into the rapid economic growth region must be prepared with local intelligence and due diligence, use local and Diaspora consulting resources, avoid stereotypes, shun pre-conceived notions, understand and engage the host community; informed bold stake-holding in this mostly untapped world is bound to yield huge returns.

4. In order for western countries to compete effectively and not be left behind in participating in the new wave of rapid African economic growth, they must do some of what China does: interest-free or liberal loans, engage host communities, invest in the future.

5. African governments must create the enabling environment for international participation in the economy: constitutional, stable government, self-reliance policy, peaceful atmosphere, anti-corruption fight, ensure freedoms and provide investment incentives. They must ensure diversification of the economy based on local commodity conversion and processing.

###

Chapter Two

PRACTICAL SOLUTIONS TO PROBLEMS OF SECURITY AND SAFETY IN NIGERIAN PETROLEUM AND ALLIED INDUSTRIES

PREAMBLE

The strategic economic importance and the highly hazardous nature of oil and gas have made the industry highly vulnerable. The vulnerability of oil brings Safety and Security problems to its industry. This paper examines various aspects of Safety and Security problems in the Nigerian Oil Industry, and discusses current and future trends in solving the problems, with particular references to the petroleum marketing sector.

SECURITY QUESTIONS AND ANSWERS

Under security, this paper will deal with the notion of security, the four aspects of security namely economic, technological, sociological and external threats of aggression.

CONCEPT OF SECURITY

The word security as used here is the generic name given to describe the state of stability and resilience of the oil industry to internal and external forces of affliction that may threaten its development or its very existence.

Since there is a comprehensive treatise on safety further below, the discussion on security here will focus on the external factors that endanger the oil industry beyond the threats of fire, and related accidents, burglary and pollution, etc., which can be regarded as internal factors.

The simple concept of the x-ray being done here is that a secure industry is one that survives and grows, bringing prosperity and peace to the people. An insecure industry is the opposite. The buffer between the two is very thin but tenuous and constituted by limiting forces or dangers surrounding the industry, that must be overcome.

Economic Factor

The broad spectrums of companies in the oil industry have a wide range of economic and financial problems which threaten their survival and thus security.

Starting with the exploration/exploitation companies such as Shell, Texaco Overseas, Agip Oil, Elf, Mobil, etc their current major economic/financial problem is that they are getting about one-third the revenue they used to get from the same capital and operations to explore new oil fields. This means the various associated service providers - seismographic drilling, cementation, logging and pipeline companies have little or nothing to do and are threatened by huge losses on idle equipment and manpower and ultimate bankruptcy. The decline in activity makes investments in these companies impossible and existing shareholders in panic, are disposing off their stocks at rock bottom prices. As these companies are mostly private corporations, members of the public are not affected. But the families of the workers of these companies who are placed on compulsory/indefinite leave, retrenchment or otherwise disengaged are certainly affected. These job seekers should, therefore, be warned that right now there is <u>labour insecurity in the oil services</u> sector of the industry, and hope may be dim.

As for the refineries, they are still in production at nearly 100% capacity. There is more than enough crude oil for them and due to their importance in the industrial operation chain, they are not denied import licence or credit lines, even under the tight foreign exchange control of government.

However, the general economic slump is seriously affecting the petroleum marketing companies such as Unipetrol, AP, Nolchem, Mobil, Total, Agip, Texaco, Elf and the Nigerian private oil companies otherwise called Independents. Due to diminishing national revenue from oil matched by increasing foreign debts, industrial raw material importation has seriously diminished. The result is the inability of many companies to produce above 30% plant capacity. Many companies without multi-national backing have virtually closed down. More and more workers are retrenched or otherwise disengaged.

The consequence is that industrial oil consumption declined by about 60% and as more car owners and transporters are forced by microeconomic reasons to park their vehicles off the road, automobiles fuel consumption has declined by about 40%. Thus there is an aggregate of 50% decline in sales revenue of the petroleum marketing companies.

Whereas the value of shares of financial institutions is high due to *high interest rates* in the inflationary economy, and agro-allied industries are waxing stronger due to greater government/practical encouragement, the value of shares of oil companies could be seen crashing on the Stock Exchange Scoreboard.

The major marketing companies still manage to make some profits – thanks to their multinational parents and trickles of import licence, the independents are stagnated or folding up, seeking amalgamation or even outright take over.

Right now, it can be said that this sub-sector of the economy is *financially insecure* or at best have limited prospect or corporate growth and capacity to improve the quality of life of the workers.

The solution to the present economic insecurity of the industry as far as the petroleum products marketing companies are concerned is *diversification*. Right now, the most attractive business line with little or no raw material input and with tremendous government support at Federal and State levels is *agriculture*.

It is pertinent here to commend the courage and foresight of Texaco Nigeria Limited and Total Nigeria Limited who took the risk of going into agriculture and have been reaping bumper harvests since they started about five years ago.

If battery manufacturing and chemical/ pharmaceutical companies with only two or three factories/branch offices nationwide can go into farming confidently, one wonders at the hesitation of a number of petroleum marketing companies with about 300 retail outlets nationwide that could serve as ready-made distribution centres for their produce.

On the choice of the specific aspect of agriculture to embark upon, since most companies have been going into actual produce farming, there are wide opportunities in *the produce/food processing sub-sector.* There is prospect in corn-milling which mechanical system can be oriented towards wheat flour substitution to feed the bakery industry which is being choked by lack of raw material. On the other hand, an oil company may take delight in the pride of being the largest *exporter* of fruit juice in the country just as a competitor takes the lead in kero or lubricant marketing. Apart from the production company as a household name everywhere its agricultural produce is consumed, the sale of the traditional petroleum products of the company will skyrocket due to the persistence of the company name sustained by the labels on the produce packages. Some skeptics in top management would still say what has agriculture got to do with us?

The answer may be found in the question, What has ITT (a telecommunication company) got to do with buying Sheraton (the world renown prestigious hotel chain) some years ago if not image boost, and near omnipresence all over the world, naturally followed by its traditional telephone business boom?.

Technological Factors

The problem of security of technology within companies is limited. The installations in a factory for instance are fixed. The idea of any staff or others stealing any components may be classified under sociological problems such as mal contentment, sabotage, etc., to be treated later. However, due to the sophistication of industries nowadays, it is very necessary for important records such as blueprint drawings, master operations catalogs and especially computer software to be jealously guarded as irreplaceable property.

The leakage of a new production technique to a competitor can unmake a company forever. The loss of the chemical formulation of a product in a fire disaster is not an uncommon phenomenon but the significant devastating consequences should be avoided at all costs.

It is recommended, therefore, that important original documents of the company should be secured in *flameproof cabinets with special security lock*.

Important records of the company should be *data-processed and magnetic tapes, discs* and so on be produced_*in duplicate* with a set securely kept in the head office and the other kept in the respective branch office or specially designated vaults. Modern surveillance equipment are now available for detecting and checking intrusion in sensitive area, tampering with equipment or important documents etc. Examples are the *microwave perimeter surveillance, infra-red closed circuit T.V. cameras for taking pictures in darkness, and counter-surveillance devices used* for detecting the presence of hidden (spy) microphones, hidden cameras, wire-tapping, etc.

As careless handling of unwanted information can be dangerous, machines have been developed for destroying unwanted information. The destruction of the information within the office of the officer concerned is the ideal because a messenger may photocopy a document for his own ulterior motive before incinerating it. To tackle this problem, shredding machines have become part of the executive office system. But now, manufacturers have gone a step further by producing machines that follow shredding with *chemical and bio-degradation*_of the information material be it paper, card cassette tapes or any material at all.

After preserving the technological information and knowledge borrowed from abroad, more than three decades of Nigerian Oil Industry experience is more than sufficient for the development of a Nigerian technological base in the industry. Government is, therefore, advised to set up a *Petroleum Technology Research and Development Council* which will use the facilities at the Institute of Industrial Research, PRODA, the Petroleum Engineering Faculties of the Universities and Polytechnics and the Petroleum Training Institute for the design and manufacture of plants locally as well as petroleum product formulation of our own. The present crippling scarcity of imported raw materials is enough justification.

Sociological Factors

The major sociological problems at company level across the industry are *fraud, espionage and sabotage.*

The frauds perpetrated by frustrated, disgruntled, over-ambitious and covetous individuals and syndicates as epitomised by the <u>alleged</u> missing =N=2.8 billion and other unaccountable multi-million naira scandals which <u>earned</u> many officials long and short term holidays in Kirikiri (prison) and other prisons are fresh in mind. These cases of fraud clearly show how the sweat of thousands of hard-working people of all cadres in the industry and the fortunes of the country can be swept into an unfathomable pit of national self-pity and misery by some unpatriotic elements and their mercenary principals overseas. The need for *tighter security* at the crude oil flow meter terminals, the records and accounts departments, the depots, the bunker vessels and even on the road tankers can easily be seen.

Espionage is not quite common yet in the Nigerian Oil Industry. In fact, it is peculiar to the developed countries where different companies spend millions of dollars on research and development and guard the results and fortunes therefrom jealously in order to excel their competitors. From upstream to downstream oil companies in Nigeria, the technologies being used are those developed and perfected in their home countries and the chances of stealing the technology of another company to improve their own locally do not exist.

However, as the petroleum and allied industry grows in Nigeria, the essence of carefully guarding chemical formulations and production techniques is assuming greater importance. Of particular significance are the different grades of lubricating oils and brands of aerosols/insecticides being produced in the country. The difference between the products of different companies is in the chemical recipes and particularly the proportions which make a particular brand stand out. Most of the major petroleum products marketing companies are now in these business lines, and they certainly have security problem to cope with regarding espionage. In the case of production technique, Unipetrol Nigeria Limited, for instance, would prefer the public to be mystified by the High-Tech In-Line Blending technique in the company's Kaduna

Lubricant Plant (which is the largest nationwide, the only one in the North and *first* of its kind in *African*) rather than hear of someone copying some process engineering details of the plant.

The most important and perhaps all-embracing anti-social act in any industry is *sabotage*. Having discussed the problems of fraud and espionage, attention is focussed here on outright destruction or attack perpetrated from inside at times with *outside instigation*. Of course the usual actors are the *frustrated*, disgruntled and unpatriotic elements. Their goal is usually to see that a particular manager or administration fails or is eliminated. They often physically put a spanner inside an engine or set traps which trigger off when someone else is on duty or no one at all is around so that their track is well covered. Successful sabotage plots have resulted in the blow up of generators, boilers, compressors and other disasters including fire which leads to plants suffering catastrophic damage, and being out of production for a considerable length of time.

Due to lack of sophisticated investigation and detective techniques on the part of law enforcement agencies, enquiries often led to the removal of managers and or supervisors in charge of the affected plant who get the blame for *incompetence* in not knowing all about his plant and bad eggs among his staff.

Before discussing ways of tackling this type of problem, one would quickly warn oil industry labour leaders to watch out for any unpatriotic elements in their membership who may unwittingly discredit their unions through industrial sabotage especially when there is official industrial action.

Although advanced surveillance systems can be used in combating crimes including fraud, espionage and sabotage, experience has shown that the cost effectiveness of such systems on their own is not impressive. The systems will be installed and maintained by human beings who for some reasons can render the systems ineffective.

In various situations in the industry, it is recommended that the reasons for anti-social behaviour among staff should be studied. Anyhow dissatisfaction, mal-contentment, frustration and other psychological problems are usually responsible. There are a thousand and one peculiar reasons that may cause anti-social behaviours that cannot be examined here. However, the most important thing is to create access to the mind of the worker or staff. The effective means of doing this is to maintain *meaningful dialogue*. By this means the mind of the workers is mirrored and their grievances can be known. Through dialogue they will be better informed too and their attitude is bound to change.

Another way of eliminating anti-social behaviours among workers is to give them a sense of belonging. Apart from recognising their status, personalities, achievements, etc., *room should be created for workers to buy up to 75%* of the shares of the company. Unless there is a case of undetected mental neurosis, no share-owning worker will contemplate destroying his own investment. Strong *sense of belonging* will foster patriotism and loyalty which will help solve virtually all sociological problems.

External Aggression Threats

The significance of Nigeria as the *giant of Africa* in terms of human and natural resources, and the champion of the emancipation of black peoples all over the world particularly in Southern Africa cannot be over emphasised. The dynamic anti-apartheid policies of Nigeria have earned her a Frontline State title over 3,000 Kms from the liberation war front. Outside Africa, the opinion and vote of Nigeria at the United Nations on international economic and political issues are much sought after by interested countries.

Nigeria's international militancy is a pain in the neck for some countries that indulge in, openly or covertly back minority rule. The consequence of that irritation to some countries is even different from other issues that directly affect Nigeria's socio-economic interests such as the repatriation of Nigerian economic saboteur-fugitives from abroad. All these acts of dynamism and pragmatism have placed Nigeria in a camp opposite to some others who feel their own interests threatened. Top on the list of the opposite people is the Apartheid Republic of South Africa, and the anti-sanction champions of Britain and the United States of America. If we are honest with ourselves, Nigeria is vulnerable to attack by any of these countries. We should learn from the lessons of others such as Libya which was devastated by American jet bombers for supporting the cause of the homeless Palestinians. As a matter of fact, a radar surveillance station, petroleum terminal and missile site were destroyed, apart from the official residence of the Libyan leader. Of particular interest here are our petroleum terminals and depots around the coastline at Atlas Cove, Bonny, Warri, Port Harcourt etc. Vulnerable in similar ways are the refineries in Port Harcourt and Warri. Of course, in a matter of minutes enemy jet bombers can reach the Kaduna refinery, NNPC depots and pipelines in the hinterlands. The experience of Zambia, Zimbabwe, Botswana and Angola – important frontline states which civilian, military and economic targets were recently attacked by the apartheid South African Air Force also buttress the reality of Nigeria's vulnerability to external *military aggression.*

A crippling form of aggression is the economic type being unleashed by the United States of America on Zimbabwe for criticising President Ronald Reagan whose policy of constructive engagement has ridiculed the US due to its abysmal failure, all economic aids are being withdrawn.

A more damaging economic aggression against oil exporting states including Nigeria is the conspiracy among the developed countries to force down oil price to its present spot market value of $5.00 per barrel from about $30.00 within a period of six months. The impact of falling oil prices on Nigeria's foreign debts and industries that depend on imports, public projects and welfare programmes is already being felt by everybody.

So how safe is Nigeria's oil industry from external military and economic aggression?

The appropriate counter-measure for military aggression against Nigeria is a formidable Army, Navy and Air Force. While the National Defence Council is the most competent authority to deal with such strategic matters, it is opportune at this moment to direct their attention to the need for establishing military units of the right mix specifically for the protection of our major oil installations throughout the country.

The military personnel must be well equipped with modern equipment especially anti-air craft missiles to counter any air raid. Early warning radar system is most important for tracking enemy crafts ever before moving close to target. The personnel need high motivation – which can be engendered by good national leadership – in order to perform effectively. In the case of economic aggression, this can manifest in denying Nigeria raw materials and spare parts for our industries or blocking Nigeria's efforts in securing external loans. There are many diplomatic-economic counter-measures that we can take against our enemies - such as nationalisation, boycotts etc. But as a nation of compulsive importers such measures cannot work. It is advisable, therefore, for Nigeria to tackle our problem from the roots that is by adopting and maintaining a policy of inward-looking development.

We should not isolate ourselves but within some target dates should produce and manufacture locally up to 80% of our raw materials, machinery, tools and spare parts needs. In short, we should reduce our dependency on foreign goods and services so that we can be in a strong position to take reprisals against any economically aggressive nation. As far as the oil industry is concerned less than 20% of our needs for machines, spare parts and other hard-ware can now be obtained locally which further under-lines the importance and urgency of the need by the Federal Government to establish the *Petroleum Technology Research and Development Council.*

SAFETY PROBLEMS AND SOLUTIONS

The problems of safety in the oil industry will be considered under the following headings: (i) Fire Safety (ii) Environmental Pollution (iii) Product Handling and Quality Control and (iv) Occupational Hazard

FIRE SAFETY

Fire Hazard

Virtually all industries and homes depend in one way or another on energy derived from petroleum products. It could be in the form of thermal power plants on the National Grid supplying electricity, boilers, furnaces, diesel engines propelling ships on the high seas or moving machines of all sorts in all kinds of industry, aircrafts, millions of petrol engine automobiles on the roads, gas lamps and gas stoves in many homes.

The process of conversion of the energy in liquid or gaseous petroleum into heat, light, sound, kinetic or other forms of energy is often accompanied by combustion which takes place under certain controlled conditions.

But if combustion or fire takes place when it is not required then disaster breaks out. From the applications of petroleum products mentioned earlier, fire incidents can occur anywhere – in the home, office, industry, highways high seas or in the air. But certain conditions must be present before fire is created. As a corollary without certain conditions or elements, fire will not be created or will be quenched naturally. What are the elements needed to create fire? Fire is the combination of heat, combustible materials and oxygen. Whenever these three elements are present at once there will be fire. Remove any of these elements and fire will die. The principle of fire prevention and firefighting is to avoid or eliminate at least one of the elements.

Sources of Fire

The first major cause of fire is carelessness followed by wilful damage and share accident. Carelessness assumes greater importance than the other causes because if it is removed, the other causes can be greatly controlled if not eliminated.

Right from the conception of a project or plant care has to be taken. Where petroleum product is being handled, extra care must be taken. Safety consciousness must go right through the design, construction, commissioning, operation and maintenance stages of the plant. This can be done through the selection of the right experts at various stages of implementation of the project. For instance an expert in huge water storage tank design or construction is not necessarily good for petroleum storage facilities whereas some ignorant clients don't know and don't bother to find out. Much as the facilities may look identical the details of high temperature, high pressure, and flame proof specifications for valves, pipes and thicker tank shell gauges, chemical corrosion/deterioration resistance, and even radiographic examination of welds distinguish petroleum installations from say water installations. The right designer, constructor, tester, and maintenance experts must be engaged, failure of which constitutes – for petroleum products – is a fire hazard waiting to happen. A trained eye and mind can always tell the difference between equipment and fittings that meet specification and those that don't.

Having got a sound plant from sound experts, one of the commonest ways of starting fires in petroleum plants is the display of naked flames and sparking devices in the vicinity of petroleum products. By their nature petroleum products vaporise easily and especially kerosene and petrol (gasoline) do so at normal atmospheric temperature and pressure. More dangerous is liquid petroleum gas (LPG) which can be in the atmosphere several metres from its leaking storage tank without being noticed. It is therefore very important to completely avoid all sources of sparks. Any inevitable or likely source of spark has to be identified and approved by the appropriate authority and given adequate protection at all times.

Other important sources of fire are electrical switchgear/fittings and diesel/petrol engines including vehicles and forklifts within the petroleum installation. For instance, the specification for electrical switches and fittings is flameproof and metal clad enclosures. For the avoidance of fire, nothing else should be accepted from the installers. In old plants, mostly old dilapidated electrical switches and worn out protective devices are found. Some of them might have been modified several times to take extra loads leading to electrical overload. The solution to this isue is electrical re-design of old plants and installation of modern maintainable equipment.

In the case of diesel and petrol engines within the installations, there are often standby generators, tanker trucks, lorries carrying packaged products, forklifts and cars of the company, staff and customers. One way or the other these contrivances cannot be avoided. But they can be controlled. The hazards they pose can be reduced to the barest minimum. Outside-the-yard parking space should be specially created for customers on whose patronage the company depends. Staff cars should be restricted to office area.

All tanker trucks, lorries and cars entering the plant must be fitted at the gate with *spark arrestors*. The spark arrestor which is a small cylinder with one closed end and several perforations on it has a screw on its side for tight fitting to the engine exhaust pipe which it protects. This locally fabricated spark arrestor has been found to be very effective and is already in common use.

But the problem, as in the case of any Nigerian rule is, *enforcement*. Many vehicles can still be seen today running up and down the plants without spark arrestors. Some company/management vehicles are also guilty of this unsafe act. It is therefore necessary to make it a punishable offence for the security man to forget or otherwise fail to fit the spark arrestor on any vehicle being admitted into the petroleum depot yard. By so doing, the security man would raise alarm at anyone refusing the fitting of the spark arrestor and higher authorities may then intervene to enforce the rule.

Similar measures should be extended to cigarette smokers and match box/lighter carriers. They must surrender their highly potent 'time bombs' at the plant gate.

The situation at the petrol (gas) station which is one of the most uncontrollable public places is equally hazardous. The petrol attendant should be empowered to refuse to sell to any motorist whose engine is running near the pump island. This also applies to motorcyclists whose worn out engine exhaust sparks have caused major disasters in fuel stations. Of course the *No Smoking* sign must now be accompanied with the *Switch Off Your Engine* sign in all fuel stations and depots.

Fire Detection and Defence

Further to the physical measures of fire prevention – that is total avoidance of sparks in flammable environment which must be strictly maintained, the prevention of the spread of fire at its incipient stage is the next important measure. The prevention of fire spread can only be achieved if the fire is detected and appropriate signals are sent out for fire defence. To this end, the *installation in plants of heat, smoke, fire, flammable vapor detectors is highly recommended.*

It is most unsatisfactory that about 80% of plants in this country only have firefighting equipment but no fire detection system. By so doing they severely limit their fire security alert level. This unsafe situation is paid for in high insurance premiums. Is it not wise therefore for companies to save high annual insurance premiums to install fire detection systems and further save the agony and cost of production time losses while awaiting insurance claims to erect new plants?

It has been noticed that the few plants that have fire detection systems can only raise audible alarm within their premises whereas the fire outbreak may be out of scale and warrant the assistance of the public *Fire Brigade*. It is therefore recommended that all installed fire detection and alarm systems must be equipped with facility of signalling over *public telephone* line or the plant *radiotelephone* in case of rurally sited plants. The signal transmission should be arranged to call the *Fire Station*, the *Police Station* and the Plant Manager.

Fire hydrants, generously distributed are a common scene in many plants. This is commendable. The fire hydrant line fed from dedicated huge water tanks for firefighting and connected to sprinkler system on and around petroleum product storage tanks is the standard practice. But it is possible for this elaborate fire hydrant system to fail when it is called upon to perform. This can happen if the *diesel fire water pump* fails to operate. The usual cause of the pump failure is dead starting batteries. For the avoidance of disaster therefore daily test run of the diesel pump must be strictly observed. There must be a float charger to keep the batteries warm at all times.

Experience has shown that only 15% of people in plants at the time of fire outbreaks respond properly and demonstrate the courage to help put out the fire. Although 15% are willing to help fight the fire, lower percentages are actually able to operate the gadgets effectively. This makes it probable that the fire water pumps will or will not be switched on when required. For greater certainty it is highly recommended that in addition to manual operation, *automatic start* of the fire water pumps triggered by signals from the fire detection system should be provided.

Effective maintenance is the bane of our equipment and plant installations in Nigeria. It is therefore not impossible for the required heat, smoke, flammable vapour detector to operate when required due to poor maintenance. This created a more dangerous situation than if there was no fire detection/alarm system at all because the mind of the people would have been conditioned to relying on the system which then fails. The solution to this problem is the installation of the *Total Environment Monitor.*

The Total Environment Monitor (TEM) which is an advanced form of the usual fire alarm control panel monitors the condition of equipment wired to it and sends an early warning signal to the appropriate location for prompt action. It is a computerised system which can be programmed to detect open circuit or short circuit of fire detectors, good or bad sprinklers too high or too low temperature, pressure, product level in tanks. These are the inputs. The output is in the form of signals to connect the telephone to the maintenance office, the fire department, the police or to operate some valves, release extinguishers, open a gate or even shutdown the plant. It can be equipped with Video Display Units or Line Printers. The system has auto-diagnostic facility which indicates audible and visibly any faulty printed board or electronic component in its circuitry. It is not too sophisticated for this reason, neither is it magic. In fact the technology is the same one found in Electronic Telephone PABX systems which have been installed and maintained in the country for more than a decade now.

It should be added that the Total Environment Monitor system can accommodate Intruder/Burglar Detection/Alarm system based on either the

ultrasonic, microwave or infrared technique –
normally used for the *day and night surveillance* of
premises perimeters that run into thousands of
metres. For total environmental protection against
fire, burglary, terrorism and vandalism, the Total
Environment Monitor which has been successfully
used worldwide to protect high risk and strategic
installations, is commended to Nigerian companies
especially those in the all-important petroleum
industry.

Safety Professionals and Fire Inspectorate

The recognition of professionalism in safety and
security matters is the first step towards a safe and
secure industry. Gone are the days of appointing
unqualified or saturated staff from other vocations
as safety and security staff. In Nigeria today there
are only two distinct institutions that offer formal
training in the safety profession. They are the
Petroleum Training Institute (PTI), Warri and the *Federal
Fire Department* and the Fire Department of a few
states. Just like there is a standard of training and
qualification for engineers, accountant, doctors and
other professionals, the *minimum* standard for the
appointment of safety staff from supervisory level

and above must be Diploma or Certificates (to ND – National Diploma, and HND - Higher National Diploma level) from the PTI or the Fire Departments.

In Nigeria no University is offering courses in this area whereas a number of institutions particularly in the USA and UK offer courses in Combustion Engineering, *Fire Engineering*, Industrial Safety, Industrial Psychology and Criminology. These are suitable courses for safety and security Managers. It is highly recommended that the private sector should send suitable staff to these courses overseas, and not wait for government (till doomsday) to establish such courses locally. Furthermore government and industry should recognise and encourage and subscribe to the Nigerian Association of Safety and Security Professionals the core members of which are trained, experienced and internationally affiliated and are in position to help set up in due course the invaluable *Institute of Safety and Security of Nigeria.*

It is also recommended that there should be a good *Fire Inspectorate* in the Fire Department which will regularly visit oil installations and depots. The

Fire Inspectorate will issue periodic certificate, recommend improvement of safety facilities or close down unsafe installations. This will be a good complement to the initiatives of the internal safety officers whose proposals never see the light of the day due to some Heads of Department who are not disciplined in the safety profession and not committed to it. It is on this note that I strongly recommend that if the Head of Safety cannot be made a full-fledged Head of Department, he should report directly to the Chief Executive who would be aware from time to time of the state of safety and security of the company and approve necessary measures for improvements.

Industry Safety Cooperation

Although there is an Oil Industry committee of Chief Executives, and there is one for Personnel Managers and Shipping Coordination, it is sad that there is no such standing committee for Engineering, not to talk of Safety. The closest industry-wide body existing is the one on Oily Waste Disposal which is not enjoying sufficient support from higher quarters.

United, they say, we stand, divided we fall. The fact that oil installations of petroleum marketing companies for instance are mostly concentrated in one area, can either be a fortune or misfortune in case of fire, depending on the level of cooperation existing among the companies. With a fire outbreak in one of the installations, if there is a standing arrangement of *Joint Safety Action*, the other companies can easily be contacted to come to its rescue.

Without cooperation the one company installation may burn down, and with only one 'good enough' explosion the fire may be extended to other installations. It is strongly recommended that an Oil Industry Safety Committee should be established to coordinate fire prevention and rescue programmes. Sub-committees for oil exploration/ exploitation companies and another for Petroleum (storage) Marketing companies should follow due to the specialised nature of the required rescue operations.

ENVIRONMENTAL POLLUTION

Human, animal, plant and aquatic life can only survive under certain environmental conditions. Whatever the differences in temperature, pressure and humidity, the most essential parameter is *clean atmosphere.*

From the diverse oil geological locations and sources, through its processing down to its multifarious uses, pollution by oil and derivatives has been one of the greatest threats to all living creatures and plants.

Oil Well Blowout and Spillage

The most devastating source of oil pollution at exploration/exploitation stage is oil well blowout whereby excessive underground pressure breaks all barriers/conventional restraining equipment and millions of litres of crude oil are discharged into the environment.

The greatest oil well blowout disaster in Nigerian history was on the Texaco Overseas Funiwa – 5 offshore rig on January 17, 1980. *146,000 barrels or 29.2 million litres* of crude oil gushed into the Atlantic Ocean. Life killing carcinogens were released in their millions. By the time the pressure subsided 12 days later, ocean waves had spread the oil to the creek areas and Niger Delta shores. The life of many aquatic creatures was destroyed. The trees in the creek areas assimilated the oily water which subsequently killed them. Of course human activities and life in neighbouring villages that depend on seafood and the creek trees were disrupted. The disaster went down as one that should never be allowed to repeat itself.

However, as man is not perfect, such accidents could still happen, so we should always be prepared.

Firstly oily waste farming prevents normal agricultural activities which can be a very serious problem where there is limited land area. The second problem is *pipeline oil spillage into surface and underground water sources* which pollution can destroy human life. Due to these problems land farming has not been a popular method of oily waste disposal.

Incineration appears to be simple but it is expensive. Energy which could be gainfully used is required in large amount to burn off oily waste. And this means money being wasted on a continuous basis apart from the capital equipment cost. Besides, the release of noxious flue gases into the atmosphere complicates the prevailing problem of air pollution. Lastly, the problem of disposal of left over ashes from incineration is created.

Re-refining appears to be an attractive way of oily waste disposal. The main problem in re-refining is the *collection and transportation to the refineries of waste oils.* There are many collection centres as there are countless users. Whose responsibility is it to set up gathering centres and organise tanker trucks to take the stuff to the refineries locally or to export? The logistic is complicated and expensive as that of distributing finished petroleum products from the refineries. Studies have shown that it is only the waste generated within the refineries and the industries and petrol stations in their immediate neighbourhood that can be re-refined economically.

Encapsulation involves blending oily wastes with appropriate aggregates of *quicklime* to form a material that is suitable for road construction - earth filling and even surface patching.

Maximum safety measures should be taken in the first instance. Standby pressure equipment that automatically takes over in case the duty one fails must be installed. In case a blowout incidence happens, it is difficult for a single company to do much about it. It is therefore recommended that a standing *Safety Rescue Committee* should be established and equipped with the most modern facilities for combating blowouts and spillage from *tanker vessels* and combating the spread of oil on the sea shores and creeks. As it is difficult to separate spilled oil from sea water or recover it from the soil in the case of blowout on land, it is recommended that the Universities and Institutes of Research should develop Carcinogen – neutralising chemicals which can be sprayed or pumped into areas of oil blowout or spillage.

Oily Waste Disposal

Oily wastes arise from the left over from the use of oil particularly lubricants, sludges from storage tanks, spillage due to poor handling, combustion by-products, etc. The greatest oily waste problem however is that of the disposal of used lubricants and storage tank sludges and spilled bitumen. If the disposal is done carelessly, animal and human life can be destroyed. There are many ways of disposing the oily wastes, namely, land farming, incineration, re-refining, encapsulation and waste-oil to diesel conversion.

Land farming involves burial of waste oils over a land area designated for the purpose. The aim is to restrict the dumping of waste oil to the given area with the hope that the waste will decompose over a number of years. But this method of disposal brings about new problems.

The application of the encapsulated material makes it attractive. But encapsulation is most suited for sludges and bituminous wastes, whereas waste lubricants constitute about 80% of oily waste problems.

The prevalence of waste lubricants is what led researchers to the discovery of the waste *oil-to-diesel conversion* process. Machines have already been developed and marketed by two British firms

Messrs. Dieselcorp and Dieselcave. The processes are however similar. Four parts of kerosene and one part of waste lubricant blended and passed through a number of filters yield five parts of diesel oil which can be used straight away in diesel engines. This has been found to be the most effective <u>and economical</u> way of disposing light waste oils with useable finished product and without any new wastes o air pollution being created.

It is pertinent to refer here to the good work being done by the Oily Waste Disposal Committee comprising Unipetrol, AP (African Petroleum), Nolchem, Mobil, Texaco, Total, Agip and NNPC (Nigerian National Petroleum Corporation). Elaborate studies have been carried out into the economics of re-refining, encapsulation, etc. which show that millions of naira is needed for the disposal of oily waste. Government should know that such huge sum of money cannot be provided easily by the private sector. Government should therefore take the lead and the private companies would follow.

The Oily Waste Disposal Committee in its tireless quest for cheap means of disposing waste oil came across the waste oil-to-diesel conversion machine which costs a few thousand naira which is small enough to install in a petrol station. As waste lubricants constitute about 80% of oily wastes, adoption of the conversion machine by the oil marketing companies will go a long way to solving our waste oil pollution problems. By so doing, the companies will be promoting clean environment in and around their depots and especially petrol stations where the largest amount of waste oil is generated. As an incentive, government should reward the stations with the cleanest environment perhaps given token tax relief, and impose fines on the dirty ones. The states Waste Disposal Boards have a role to play here too.

As in Ogun State, government should establish *mechanic villages* which help restrict to limited area broken vehicles, related hygiene problems and waste oil generation. The State Waste Disposal Boards and the local government could buy the conversion machines locate them in the mechanic villages and re-sell the process product of diesel to the mechanics. Revenue will be raised and a clean environment will be maintained.

Finally the Environmental Protection Agency decree which has been on the drafting table for many years should be promulgated. It will give strong backing to the existing state edicts and bye-laws on pollution. It will establish on full-time basis the policing of our environment, promotion of education, research, and establishment of safety standards for the environment.

PRODUCT QUALITY AND HANDLING

The incidences of kerosene explosions leading to deaths across the nation in recent times constitute the greatest disaster of the oil industry that had direct and immediate impact on the Nigeria public.

After a series of investigations the cause of explosive kerosene was traced to off-spec, dual purpose kerosene. The off-spec characteristic was in turn traced to two likely sources namely, poor quality control at the refinery, and contamination with highly flammable product especially petrol (gasoline) due to poor handling.

Our refineries are few and are known to have qualified and experienced staff doing a good job over the years, so it is quite easy to eliminate any problem from their end. And if any quality problem is traced to ineptitude or negligence at any of their laboratories the consequence can be severe and the persons responsible can be identified fairly easily and disciplined.

On the other hand the problem may lie between the gates or distribution jetty of the refinery and the consumers. The product changes hand about four times in the case of kerosene; first on to the tanker vessel, then to the marketers storage tank, then to the tanker or peddler's truck, then to the reseller's skid tank, then to the customer's kerosene jerry cans before the stove. The likely sources of contamination are: (i) the product pipeline from the receiving jetty to the marketer's storage tank – if the pipeline which may be about 2km has just been used for receiving petrol, (ii) the tanker truck which may have been used for petrol by a previous shift plant man (filling operator) under pressure of truck shortage (iii) the consumer's kerosene jerry can might have been used for buying petrol the previous day. The detail analysis of circumstances that could justify any of the above likely causes cannot resurrect the dead victims neither is it the theme of this section of this paper. The scenario of *death from kerosene explosion* which can happen to any of us during simultaneous (NEPA) utility power failure and gas shortage, or our extended families who depend daily on kerosene, is meant to focus our minds on the grave responsibilities on our hands for High Quality Control and Efficient Product Handling.

The measure of distinguishing kero trucks from petrol trucks by *colour identification* and strictly maintaining them for carrying particular products only has gone a long way to eliminate the chance of contaminating the products through indiscriminate product loading. In the installations, product lines must similarly e *colour-coded*. The main pipelines from the jetties must always be flushed with diesel before any flow operation, and the flushed product and interface must be passed into a slop tank. The NNPC's billboards *educating people* on proper handling of kerosene is highly commendable. The same medium can be used to further warn kero users with incisive <u>slogans</u> such as 'It kills to use your petrol can to buy kerosene', or 'Only dead men and women use petrol can to buy kerosene'.

The damage done by poor or wrong lubricants is also devastating though only to the engine concerned and of course the purse of the owner. There were numerous cases of knocked engine between 1983 and 1984 when *adulterated engine oil* flooded the market. Thanks to the Ministry of Petroleum for stamping out road side lubricant dealers. There was recently a case of engine oil found in the rear axle of a made-in-Nigeria car. Did that come from carelessness or by accident?

At the lubricant plants, the use of *wrong labels* on product cans or kegs - especially in these days of scarcity of plastic raw materials among other things - can ruin many engines and damage the reputation of marketers.

High toxicity of aerosols is hazardous, and erroneously filling air freshener cans with herbicides can be costly. It can kill.

All the above, point to the fact that petroleum companies, particularly marketers with storage handling facilities must maintain an up to date laboratory for quality control with *periodical inspection/certification.* The industry quality assurance superintendent companies must maintain high laboratory analysis standards and professional ethics. The *consumers must be educated in local languages* too <u>on</u> the safety aspects of the products. A strong *Consumer Protection Association* must be established possibly under the aegis of the Public Complaints Commission and certainly with the assistance of the Press, that is a fair, investigative and thorough Press.

HEALTH HAZARDS

Sources of Hazards

The occupational hazards in most industries with regard to ergonomics - working conditions of temperature, pressure, humidity, standing or sitting position/duration – are almost the same. What is particular to the petroleum industry are health hazards which are discussed herein.

Due to the volatility of petroleum products, most of them vaporise at normal atmospheric temperature. Consequently the working and handling environment has a high potential of being laden with petroleum products vapour which can affect human beings, even without physical contact.

Due to their toxicity, exposure to the vapours or real physical handling pose a lot of dangers to the workers and users. In order to emphasise the seriousness of the health hazards of the petroleum products the chronic long term case of exposure and contact will be treated briefly.

Various chemical elements and compounds are found in crude oil, the processed base mineral oils and the additives which give different properties to the products depending on their applications. Some of the elements and compounds are naphthalene, paraffin, lead, olefins, esters, silicones, ethers, alcohols, organic acids, amines, zine, calcium, magnesium, thiophosphates, phenates, triazoles, triazines, sodium, and chlorine. By exposure or contact the chemical elements and compounds affect living things with devastating results. The simplest illustration is the effect of blowout or spilled oil in creek areas which affect mangrove trees by killing the chlorophy1 in the green leaves, turning them brown, dropping them off and eventually killing the trees – an ecological. Due to the intermittence of similar exposure and contact by man, the reaction is slower but over a long time the result is the same.

Effects of Hazards

Exposures to or contact with petroleum products may affect man through the skin, the eyes, through inhalation and ingestion.

(a) <u>Skin</u>

Skin rash and oil acne – recognised by the presence of blackheads, pimples and pustules – may occur at the site of exposure or contact when exposure is repeated and prolonged in workers with poor personal hygiene. With some poorly-refined base oils, warty swellings or sores may ultimately develop which do not heal and may lead to cancer.

(ii) <u>Eyes</u>

Repeated exposure of the eyes to mineral base oils will lead to eye irritation.

(iii)<u>Inhalation</u>

Prolonged and repeated exposure to significant atmospheric concentration of mineral oils may lead to a benign form of lung fibrosis, possibly preceded by symptoms of bronco pulmonary disease.

(iv)Repeated accidental ingestion is unlikely to occur except for mechanics that use their mouth to bleed fuel in motor vehicles. The secondary ingestion due to inhalation of oil mist is however so small in actual volume that effects such as the irritation of mucous membranes are not expected.

Precautions

Environmental Control – keep uncontrolled and undesired releases of lubricants to a minimum by correct oil selection and application, by fitting splash screens or by enclosing machines or parts.

Work Method
Physical touch of petroleum products should be avoided as much as possible. Avoid the use of compressed air for cleaning oil, instead use rags which in turn should be hung up and not kept in overall pockets. Suitable gloves, aprons, overalls, masks, and footwear in various combinations can effectively minimise contact.

Personal Hygiene
Adequate washing facilities, clean work clothing, laundry facilities, barrier creams, footwear, locker and changing rooms are highly recommended.

Information
Workers must be advised on the possible harmful effects of the products they are handling to give them necessary safety consciousness. Medical supervision twice a year is recommended. Workers should also be aware of the provisions of the law and their rights. Consumers should also be informed through packaging labels and through salesmen. Warnings on exposure to insecticides are a must.

Management and Union
Management should make adequate budgets for all the above. And the labour unions armed with the provisions of law, the necessary medical services, should monitor the situation in their respective work circumstances and ensure that management does not shirk its responsibilities; neither should workers make unreasonable demands.

CONCLUSIONS

The conclusions on the Practical Answers to the Question of Security and Safety in Nigerian Petroleum and Allied Industries, with particular reference to the petroleum products marketing sector are presented in form of checklist of essentials for security and safety, as follows:

1. Business *diversification* of marketing companies will guarantee *economic security* and prosperity with agriculture being highly prospective now and petrochemical downstream industrial ventures in due course.

2. Preservation of important plant specifications, technical drawings, product formulations using data processed tapes and discs or microfiche where applicable is very essential. It is advisable for government to set up a *Petroleum Technology Research and Development Council* for the promotion of design and manufacture of plants and product formulations locally. India has done this successfully and we can do the same.

3. Sociological problem of *fraud and sabotage* are created by frustrated, disgruntled and over-ambitious workers. It is necessary for managers to be trained in Industrial Psychology to fish them out before any harm is done. Management should however ensure congenial working conditions and delightful welfare schemes are maintained.

4. Threats of external aggression can be contained through the provision of crack defence units especially for the protection of our major oil installations particular from air raids. Deliberate policy of self-reliance should be pursued by government in the oil sector to counter any economic aggression by any country trading with Nigeria that may apply economic sanctions against us, just as the US is doing against Libya now.

5. For fire safety it is essential for petroleum installations to be adequately provided with flammable vapour detectors, heat detectors, smoke detectors and alarm systems with facility for automatic contact with local fire stations. The introduction of the *Total Environment Monitor* will give early warning of any fault in the fire detection and defence system.

6. The establishment of an Oil Industry Joint Safety Action Committee is imperative for co-operative efforts in combating fire and similar hazards, it is recommended that safety should be given due recognition as a distinct and responsible profession. For maximum effectiveness the Head of Safety be trained to

at least HND level on courses such as Fire Engineering, Industrial Safety and Criminology – and should report directly to the Chief Executive. An effective government Fire Inspectorate is most essential.

7. For environmental pollution control, the Federal Government should make substantial financial contribution to the private sector's efforts in oil waste disposal through *re-refining and encapsulation.* The use of oily *waste-to-diesel conversion* machine by machine workshops, mechanic villages, petrol stations or the other petroleum marketer's installations, should be encouraged. Government should allow the importation of the easy to operate machine duty free to alleviate the scourge of waste lubricant pollution; followed by local manufacture

8. The establishment of *Oil Consumers Protection Association* under the aegis of the Public Complaints Commission is recommended for the protection of Oil products users of all classes against contaminated products (like kerosene and engine oil in the past and other devastating industrial negligence.

9. Necessary precautionary measures should be taken by oil users to avoid health hazards through product contact with the skin, the eyes, inhalation and ingestion by the use of personal protective equipment - gloves, aprons, goggles and footwear. Work barriers, clothing, laundry facilities, barrier creams, locker and changing rooms must be provided. The above facilities including routine medical check-ups must be adequately funded by Management assisted by the vigilance of the labour unions but within sound medical and legal bounds.

10. Consumers and workers education on the dangers to which they are exposed, will go a long to prevent the hazard. Packages of products such as insecticides must carry warnings on the hazard of exposure.

###

Chapter Three

MANAGEMENT ACCOUNTING
A Planning and Decision-making Toolkit for
Small and Medium-Scale Businesses

THE ESSENCE OF MANAGEMENT ACCOUNTING

In order to grasp the essence of management accounting and adapt its value to small and medium scale businesses, it is necessary to understand different concepts, types, and purposes of the accounting function and reporting.

Definitions - Management Accounting is the provision of accounting information to assist all levels of management in the function of planning, decision-making and control. The classification of accounting given below shows the distinction between the main aspects of management accounting and traditional financial accounting.

Stewardship Accounting – Profit-and-loss accounts and balance sheets.

Decision Accounting – Estimates of costs and revenues associated with particular alternative (make or buy or hire).

Control Accounting – Information to assist management in measuring and minimising variances that adversely alter plans.

Whereas the professional accountant is primarily concerned with the preparation and auditing of the Profit-and-Loss Accounts and the Balance Sheet, he or she is merely dealing with the account of stewardship of the previous year. The Management Accountant on the other hand is busy with extracting and presenting data to enable future decisions and control of the business. The table below compares and contrasts Financial and Management account for clearer understanding.

Factor	Financial Accounting	Management Accounting
Purpose	External consumption to meet Tax requirements, Companies and Allied Matters Decree requirements for the Registrar, Shareholders and Public needs	Internal Consumption for management to serve an economic purpose greater efficiency and higher profitability
Period	Concerned with the past	Mostly present and future
Frequency	Final Accounts and Balance Sheet-annually	Routine reports-monthly, weekly, and daily.
Promptness	Within 2 to 6 months of accounting period	Prompt production
Accuracy	Absolute accuracy required	Approximation acceptable
From	Prescribed by Companies Decree which sets minimum content.	No set form, designed to suit individual requirements.

Fig. 1: Financial and Management Accounting compared

Our focus will be on processes in planning, decision and control accounting.

Decision accounting – is concerned with the evaluation of alternatives involved in short term and long range plans. It may be a decision such as investing in additional buildings, opening a new branch, the introduction of a new product, the choice between two contracts where there is limited plant capacity or other resources, the altering of the selling price of oil, etc. The nature of cost comes into consideration as well as time factor. Various aspects of management decision, time value of money will be examined.

Management Control accounting – is based on planning and standards followed by the measurement of departure of actual performance from set standard and plans. Once measured, the variances are analysed as to causes, the vital ones of which are then tackled to keep the enterprise on course. Management accounting assists management in achieving cost control and cost reduction to meet its aim of profit-making.

Besides the skills of decision and control accounting, the effective management accountant must have good knowledge of his business and human environment for survival and success of the business. Cost is a critical driver in business operation. It is important for the businessman to understand costs and costing in day to day operation.

COST CONCEPTS

Introduction - The principal functions of the cost accounting are to record, classify, allocate, apportion, collate, summarise and report current and future costs to management.

For the purpose of planning and control, cost data could be used by management as

(i) Communication device - stating plans, objective and limits
(ii) Motivation device - incentive for management and workers agreement and action
(iii)Appraisal device - reference point for performance measurement.

DEFINITIONS AND CLASSIFICATION OF COSTS

Cost Unit - unit of product, service, time to which cost may be traced.

Cost Centre - location, person, or equipment for which cost may be ascertained.

Cost Elements - various components of costs.

Direct Cost - one that can be allocated directly as a whole to a cost unit or centre.

Indirect Cost - one that cannot be directly associated to cost unit or centre but apportioned, e.g cost of lighting.

Cost Allocation - allotment of whole items of cost to cost unit or basis.

Cost Apportionment - sharing of cost items among cost centre on suitable basis.

Material Cost - cost of commodities supplied to cost centre and consumed by cost unit.

Labour Cost - cost of remuneration (wages, bonus, commissions, etc.) of the employees of an employee.

Expenses - cost of services provided to an undertaking such as water, electricity, and rational cost of asset consumption (depreciation of plant).

Prime Cost of Production – the sum of all direct costs and direct expenses.

Overheads - the sum of all indirect cost of production

Absorption of Overhead - achieved by the use of one or a combination of overhead rates e.g. labour hour rate, machine hour rate

FIXED OVERHEAD

Costs - tend to remain at the same level irrespective of the volume of production, but accumulate with the passage of time e.g. rents, office salaries, insurance.

Time Cost - or fixed overhead costs – accumulate with the passage of time.

Variable Over - tend to vary with changes in volume of production but head

Costs - cannot be allocated directly to the unit cost e.g. packaging.

Semi-Variable - are partly fixed and partly variable e.g. electricity to start Costs up machines (fixed), electricity consumed as production volume rises (variable). In the long run costs do change.

Variable Costs - all prime costs are variable, being traceable to cost units

Production Cost - sum of prime costs and production overheads.

Selling and Admin. - indirect cost such as salesmen's salaries, commissions, Distribution.

Overhead - advertising costs, office salaries, office rent, R&D, depreciation, etc.

Total Cost - sum of all prime costs allocated and all overhead costs apportioned to a cost unit.

Seasonal Costs - vary with seasons irrespective of volume of production e.g. heating in winter.

Controllable Costs - are within the control of the departmental manager e.g. efficient use of labour, energy, time, material can bring down costs.

Non-Controllable-able Costs - are outside of the departmental manager e.g. rent of factory premises, which cannot be used to judge his performance.

Relevant Cost - cost that have impact on (future) decision, past costs are irrelevant to decisions.

Sunk Costs - past costs which will not alter future decisions e.g. cost of existing plant, depreciation, etc.

1. Direct Materials
2. Direct Labour
3. Direct Expense
4. Works indirect or factory overhead
    a. Indirect Materials

        b. Indirect Labour

        c. Indirect Expense

5. Administration <u>overhead</u> comprising

        a. indirects (material, labour, expense)

        b. applicable to admin.

6. Selling and distribution overheads

        a. comprising indirects (material, labour,

        b. expense) applicable to selling and

        c. distribution.

7. Plus – Profit Margin

        a. or Less – Loss suffered

8. Total Price (selling)

Fig 2 Structure of Costs selling price

TYPES OF COSTING

Costing is the process of calculating costs. It is an activity undertaken overtly or covertly in every commercial transaction. Whereas revenue (income) is easy to see and understand, costs show up in dozens of ways obvious and hidden, and could be complicated. Costing enables a conscious effort to plan and put cost in its place. Different types of cost apply to different situations.

Job Costing – involves the accumulation of costs for specific job, product, equipment, etc. Costs are directly associated with the job and recorded in terms or materials, labour and proportion of overheads attributable. Example: ship building, construction, printing.

Batch Costing – rather than a cost unit, a batch or products or jobs has its costs recorded. Job and batch costing are comprehensive. They however involve a lot of clerical work and could be cumbersome.

Process Costing – involves accumulation of all costs concerned with the production process over a given period of time. The average unit cost then be calculated at the end of the period. The advantages are less clerical work and less expense in this Regard. The disadvantages are the unit cost cannot be determined until the end of the given cost period and where joint products are obtained at different stages the computation of attributable costs is often very difficult. Process costing is applicable to homogenous product industries e.g. cement, sugar, bottling companies.

Historical Costing – involves accumulation of actual costs after the costs are incurred. The demerit in historical costing, per se, includes lack of clarity because of many unknown quantities, leading to lack of accuracy, as well as lack of reference standard for ensuring efficiency of operation. Hence historical costing is not used in isolation but in conjunction with standard costing.

Standard Costing – accumulation of predetermined costs. In practice, the actual costs at different production stages are compared with the standard. The differences, known as variances are analysed, causes identified and inefficiencies communicated to the responsible manager. e.g. material price variance (to buyer), material usage variance (to production Manager).

Absorption Costing – the system of costing whereby all *production overheads* are absorbed into the cost unit on a suitable basis. The basis of overhead absorption or apportionment could be labour hour rate, machine hour rate or material cost percentage, depending on the type of products and nature of overheads. For example power may be apportioned to departments according to floor area. Cost of canteen, or heating may be apportioned to departments according to personnel population enjoying the facility. Maintenance apportionment cost may be limited to production departments only. Thereafter the apportionment to cost unit may be done on machine hour rate or labour hour rate, etc, that our overhead incurred is more than calculated cost, then we have under absorption, and vice versa.

Differential Costing – when choosing between alternatives, the computation of the effect of certain line of action on costs and revenues is referred to as differential costing. In other words, that is the effect of different situations on costs and revenues. Differential costing deal with relevant future costs future costs not residual or past costs. Differential costing is a decision-making concept. It involves the task of gathering the relevant costs, and could be very difficult to apply to absorption costing systems. Differential costing can be applied to short run or investment decisions (including time value of money and discounting).

Incremental Costing – this is the type of differential costing involving changes in volume of production. It is applied to marginal costing.

Variable Costing – the ascertainment of direct or prime costs and all variable overheads (for the purpose of focusing on controllable costs)

Marginal Costing – this is the costing technique applied to the analysis of effects or changes in volume and type of output in a multi-product business. Marginal cost is the aggregate of cost increasing or decreasing the production volume by one unit at a given level of output.

Marginal Costing involves determining how an extra one unit of output will increase costs (prime costs and overhead costs). Marginal costing is used to decide on contract production and pricing. Of course sunk or fixed costs are ignored while relevant future costs only are considered.

Contribution – this is the difference between the selling price of the various products and their variable or marginal cost. Contribution represents what (contribution) each product makes towards the recovery of fixed cost and profit. In a business the line *product or services with the higher contribution* should be emphasised. This does not mean the other products should be eliminated. However detailed analysis will expose which product is not contributing to profit nor fixed cost recovery as well as customer associated-product buying habit to justify elimination of suspect product.

Limiting Factor – every organisation has a limiting factor against its activities at every point in time, e.g. sales potential, raw material, skilled labour, cash or floor space. It is vital to determine which products will optimise the company's benefits within its limited capacity. Linear programming is useful in limiting factor analysis.

Opportunity Cost – this is the cost of taking one line of action against another. It is the cost of the best alternative foregone. If investment option A attract N20m while B attracts N15m. The opportunity cost of taking B rather than A is N5m.

For example, there is opportunity cost in saving shareholders dividend to finance additional plant.

BUDGETARY CONTROL AND THE HUMAN FACTOR

Introduction - Management is about effective utilization of resources – men, material and money to realise set objectives. In order to fulfill this responsibility good managers plan very well in carrying out their activities or businesses.

The aim of business enterprise – small, medium or large - is to make satisfactory profit at minimum cost. Thus for a given target revenue or profit, there is a target or limit of cost to be incurred. This is the crux of budgeting. It is normal practice at the end of a given operating period to compare actual performances with planned activity. This is the principle of budgetary control.

The budget per se is a compilation of estimates of inputs and output, or costs and revenue. It represents facts and figures. On its own budget has meaning. It is the human involvement, and implementation of budget that makes it alive and meaningful.

The Control Function - In performing the control function as it concerns budget, management end up doing the following.

- Communication – statement of corporate intention, proposed plans, etc.
- Motivation – getting people involved in setting sectional and personal performance objectives and standards, also involving all levels of management in the sequential compilation of estimates of costs and revenue.
- Measurement of performance by comparing actual, with set objectives.

Communication, motivation and performance are elements of the human factor which pervades the whole budget process. Through budgetary control, periodic budgets are established which define the *responsibilities of managers* in financial terms. Periodic *evaluation* is then done to measure deviations from the plan and steps taken for revision adjustments to ensure that corporate policies and objectives are realised.

The Human Factor - Being that managers are individually responsible for keeping within the agreed targets, they must of necessity be involved in setting the financial limits. This is the motivational function of budgeting. It assigns responsibility and creates a sense of responsibility, commitment, and a sense of belonging. Since the manager's performance measurement and reward depends on it, the spirit of departmental or divisional competition further ensures good corporate results.

The Budget Process - The essential requirements of a sound budgetary control system are:

- Clear organisation structure and adequate accounting records
- Revision of budgets at prescribed intervals
- Strong management commitment and regular staff education on budget preparation and usage.

Budget Committee is normally formed to carry out a central compilation, coordinating and consulting function under the guidance of Budget Accountant or Officer. The budget officer receives budget performance reports, collates and analyses the deviations and recommends courses of action to be taken.

The development stages involved in the preparation of budget for a typical manufacturing outfit are:

- Determination of Basic Assumption – Corporate objectives: Profits, market share, etc.
- Forecast of the general economic conditions in the environment
- Preparation of Budget proper in sequence:-
- Sales Budget (targets based on forecast and company capacity)
- Production Budget (planned materials stock, labour, machinery availability)
- Administration Cost Budgets
- Selling and Distribution Budgets
- Capital Expenditure Budget
- Cash Budget
- Pre-payment an d Accruals Budget
- Research and Development Costs Budget
- Revenue, Funds Flow Budgets

- Profit and Loss or Master Budget.

Zero-Based Budgeting - This system of budgeting is opposed to the popular method of incremental adjustment of previous levels of costs and revenue, to derive the next period budget. Zero-based budgeting rather assumes that every new year or period and its budget starts from a zero base, and all on-going activities and projects for the period must be justified as if they were new. It examines and calls to question the need or otherwise to continue or discontinue a project rather than merely adjusting year in year out as a matter of routine as in traditional budgeting.

An advantage of zero-based budgeting is the cost-benefit analysis it forces on managers in the face of competing needs for limited resources. Everything has to be justified afresh. Deadwood projects are dropped while the profit spinners are preserved. Inefficiencies are eliminated, and allocation of resources improved rather than merely reducing cost and increasing revenue as a ritual.

Changing circumstances are also objectively reflected in zero-based budgeting. What was justified yesterday, will not necessarily hold, so why the traditional historical adjustment?

Zero-based budgeting motivates managers to seek improvements in their methods and helps to articulate the innovations for budget presentation.

This system of budgeting is much more tasking. It is most useful where greater discretion is possible over spending pattern e.g. personnel, research and development, etc.

BREAK-EVEN ANALYSIS

Definition - Break even analysis is the pictorial illustration of changes in costs and profitability at changing levels of output. The break-even chart is the basis of this analysis. It shows fixed cost, variable cost, total cost, sales break-even volume, break-even sales, production capacities of volume, loss and profit. Refer to the Break-Even Chart.

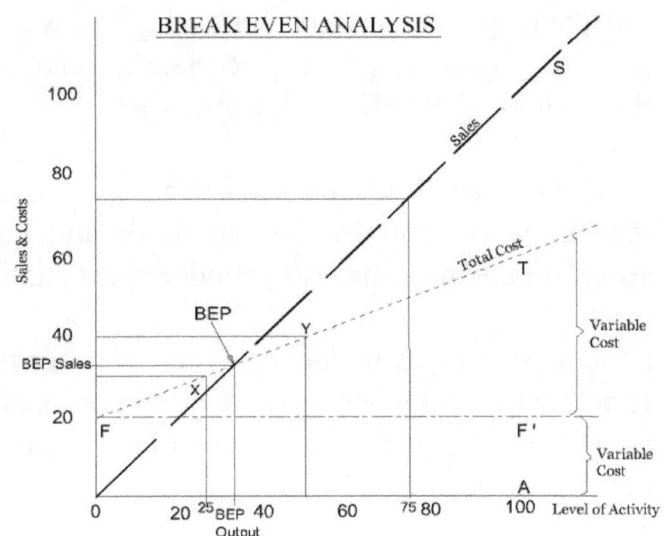

OA - Level of activity, prod. capacity, or number of units
OB - Costs & Sales in N,000
OF - Fixed Cost
FT - Variable Cost
OS - Sales Revenue
BEP - Break Even Point
TFOA - Total Cost i.e. Variable + Fixed costs

Fig.2: Break-Even Analysis Chart

It is assumed that all the items produced are sold.

It is important to note that at any point in time within the period, the business is not incurring fixed cost of =N=10,000 over the period for which the chart is drawn. Assuming 100 units are produced during the period, with total variable cost of =N=40,000, i.e. =N=400 per unit.

At 25% -   i.e. 25 unit per period – total cost given at 'x' which read off on the vertical scale is shown to be =N=30,000 (Fixed cost =N=20,000 + variable cost 25 x 400 = =N=30,000)

At 50% -   at the rate of 50 units per period – total cost is at 'Y; which read off on the vertical scale is seen to be =N=40,000 (Fixed cost =N=20,000 plus variable cost 50 x 400 = =N=40,000)

At 75% -   at the rate of 75 units per period – total cost at 'Z' read off on the vertical scale is seen to amount to =N=50,000 (Fixed cost

$$=N=20,000 + \text{variable cost } 75 +$$
$$=N=400 = =N=50,000)$$

Line O-S is the sales revenue 'curve'. In the example, the selling price is $=N=1000$ per unit to give nil revenue at nil activity, $=N=100,000$ when 100 Units are sold $=N=33,400$ for the sale of 33.4 units when THE TOTAL REVENUE IS JUST EQUAL TO THE TOTAL COST I.E – THE BREAK-EVEN POINT (bep)

Shown at p. sales $= =N=33,400 = $ Fixed costs $=N=20,000 + $ variable cost
$33.4 \text{ x } =N=400 = =N=33,400.$

Below p. production is at a loss represented by XE

Above P. production is at a profit

At P i.e. BEP, profit (or loss) $= 0$

$S = FC + VC + FC + P = VC + C$ [contribution]

$S - VC = C$

Contribution $C = S - VC$

At BEP, $S = FC + VC$ i,e $P = O$

i.e. $S - VC = FC$

$$C = FC \text{ at BEP}$$

Example $C = S - VC = N1000 - =N=400= N=600$

i.e.

Contribution per unit

Since Fixed Cost FC = N20,000 =

contribution/unit x No. g units

No of units = $\underline{FC}$ $=\underline{20,000}$ =33.3

  Contribution/Unit   600

Thus at BEP, No of Units = $\underline{FC}$

                      C [Contribution/unit]

the, sales   = $\underline{FC \text{ x Sales/Unit}}$

                  C Contribution/Unit

Where Fixed Costs are high, Break-even point sales is relatively high, thereafter profit accrues rapidly. This means great loss in depression, great profit in time of boom. Where Fixed Costs are low, breakeven point sales is relatively low.

CASH BUDGET

In order to fulfil its objectives, a business must have sufficient funds for its liabilities and proposed investment. Where there is inadequate cash for the day-to-day running of the business, its activities salaries, temptation of retrenchment and the reality of illiquidity tending to bankruptcy. Where there is too much cash available, it leads to unutilised assets. A wastage to be avoided. In a similar manner as shortage of funds, idle assets or funds will fail to generate the required returns for its owners.

Consequently, the manager in charge of finance should always have right amount of cash at all times. The key is a good cash budget on weekly or daily basis. The cash budget contains the forecasted cash inflows and outflows. The factors to be borne in mind in developing cash budgets are:

a. Credit policy of the company-say, 30 days maximum period for payment for goods sold above certain amount and cash for certain minimum percentage.
b. Debit management policy – material bills to be settled in 45-day period, quarterly

payments [energy], annual [rents], monthly payments [salaries], daily payments [casual labour], etc.

Where loans require securities and are suitable for medium and long terms purposes, overdraft facilities are suitable for meeting short term cash requirements. Cash budget should cover over draft interest and capital repayment schedule.

MANAGEMENT INFORMATION SYSTEMS

It is understood that Management accounting is the presentation of accounting information to enable management functions of planning, decision-making and control. The set of information and formats for this purpose is known as **management information system** to ensure that the relevant financial information is made available to relevant managers in the appropriate form.

Strategic Information - Strategic information is used at the senior levels of management to plan the long-term objectives of the organisation and to assess how these objectives are being achieved in practice, e.g. Level of profitability, division performance measurement, merger and acquisition strategy, purchase of major new fixed assets, raising new long-term funds and selection of major products and markets. Both the internal and external environments of the organisation must be properly considered. The accountant must therefore assess the impact of political, economic, sociological and technological, (PEST) factors in financial terms on the following:

- Projections of future interest and foreign exchange rate
- Projected levels of taxation
- Costs of raising new funds
- Projected capital expenditures and cash budgets
- Competitors' actions
- Future staffing levels and associated costs
- New technologies

Strategic information is normally produced at irregular intervals. It is broad in scope and imprecise. It demands a high degree or intuition and judgement power of the recipient managers.

Tactical Information - This is used by middle management to ensure that the available resources are used in the most effective and efficient manner to achieve the objectives laid in the strategic plans. Most tactical information is generated internally and the accountant will supply information concerning:

- Analyses control and variance analysis reports
- Short-term cash flow forecasts
- Debtors analyses
- Departmental operating efficiency reports
- Specific management requests

Tactical information is less broad, more detailed and produced more regularly than strategic management.

Operational Information - This is the lowest information level in the organisation. It is used by first rank managers, supervisors and foremen in the day-to-day operations to ensure that specific tasks are planned and carried out effectively. The accountant must provide operation personnel with relevant financial data to improve on operating cost.

Examples of operational information concern the following:

- Regular reports on the costs of wastages
- Analysis of job costs
- Labour and machinery efficiency reports and cost implications
- Detailed sales analysis by product, market, etc.

Operational information is more precise and frequently produced than others. Generally what is to be borne in mind is the provision of Good Information, which is information that allows an executive to maximise his decision-making capacity to the benefit of the organisation.

PRICING STRATEGIES

Pricing policy is one component of the marketing mix which must be integrated with the other components to meet the requirements of the corporate marketing strategy.

The choice of pricing policy depends on the following:

- Manufacturing cost of the product

- Nature of competition

- Life-cycle of the product

- Branding advertising and sales promotion back-up

- Quality of the product

- Market share objective

The ultimate objective of pricing is to recover all the associated costs and provide an acceptable profit. However other considerations must prevail in the short run. The role of the accountant would be to ensure that the marketing manager is fully aware of the financial results of his choice of strategy.

Penetration Pricing - This is a low priced strategy used in introducing a new product or service to an existing competitive market. For a reasonable length of time this policy will discourage potential competitors from entering the market. It is suitable strategy for a product that can be mass produced and for which demand is highly elastic.

Market Skimming - This is a high price strategy in which the new product is launched at a high price backed up by heavy spending on advertising and promotion campaigns. As the product becomes established and associated costs fall the price is progressively reduced. The aim of market skimming is to earn high profits and recover costs early in the product life cycle. It runs the risk of more competitors entering the market.

However it may be more beneficial for an entrepreneur in small or medium – scale business to adopt a policy of initially setting lower prices to obtain higher market shares with improving contribution, and discouraging competitors than from a high priced strategy aimed at a rapid recovery of development costs.

'Cost Plus' Pricing - Traditional pricing is on cost plus basis. The price is set by adding a profit mark-up to a cost figure which might in practice be the full absorption cost, the production cost or the marginal cost or the product. The policy is suitable where the company is a market leader and the competition generally matches the price set.

Existing Product Competitive Pricing - A market leader normally exists and a company can decide to price its products at a slightly higher price than the market leader and rely on quality and advertising to maintain a small acceptable market share. The market leader is likely to ignore this unless his market share shrinks severely.

At times an organisation spends a lot on advertising promotion to support a brand name which will attract a price premium for its products above others or preserve a higher market share. The accountant should analyse the situation to confirm that the advertising expenditure is justified. Companies at times seek to dominate the market by setting aggressively low prices. This is risky as costs may not be fully recovered while ensuring price war cripples the business rapidly.

Contract Pricing - In situation of surplus capacity the management accountant may advise the marketing manager of the marginal costs of supplying the product. This will enable specific contract orders to be sold at below normal prices to generate additional contribution which the idle capacity cannot generate. The company should be conscious to see that the company's normal business is not trapped in this low-price hole.

In general the accountant has the role of supplying financial decision to guide the choice of suitable strategy.

A DECISION-MAKING CASE:

Merger & Acquisitions

The partners in the Firm, ABC Consulting Petroleum Engineers, established in 1970 are preparing for negotiations for merger with the management and building services consulting firm of XYZ Associates Nigeria which was set up in 1980.

You are required to:

(i) State the major reasons why companies seek growth by means of merger and acquisitions.

(ii) Outline the information which the Management Accountant may provide to aid the merger negotiation between the two firms.

SOLUTION

(1) (a) Expansion of facilities, attraction and retention of managerial talents.

(b) Growth by merger or acquisition saves time and minimise risk.

(c) Diversification reduces over - dependence on one line of activity of product.

(d) Smoothens out seasonal product or service demand pattern.

(e) Reduction of production or operating costs, advertising costs, etc

(f) Increase in capacity and opportunities.

(g)   Extension of market through additional branches.

(h)   Elimination of competition and enhancement of market leadership potential

(i)   Purchase of assets at below their true value, improved asset security, etc.

(ll)   Mergers and acquisition policy of an organisation is a major plank of corporate strategy in response to PEST (Political, Economic, Social and Technological] factors and SWOT (Strength, Weakness, Opportunities and Threats) analyses.

The Accountant should pay attention to these factors

(b) Determination of the price to be paid for acquisition or terms or merger.

(c) Gains synergy of merger should go to the acquiring company while gains in shares goes to the acquired company

(d) Method of payment of the price should be considered: exchange of shares, cash or debenture loan stock.

(e) Effect on consolidated balance sheet, profit and loss account, earning per shares, gearing, etc.

(f) Retention or improvement of earnings per share, price and dividend policy.

(g) Evaluate and justify cost reduction.

(h) Examine staff number and salary rationalisation.

(i) Consider alternatives, if any.

###

Chapter Four

ENGINEERING AND TECHNOLOGY FOR
NON-OIL EXPORT
Challenges, Strategies and Expectations

ABSTRACT

The vagaries of the export market of Nigeria's monolithic economic commodity, petroleum (oil) yielded a colossal decline in earnings from $21b in 1981 to $5b in 1988. The result is a chronically ailing economy and social frustrations.

In order to turn round the economy, in 1986 government introduced incentives for intensive diversification into non-oil exports. The contribution of non-oil exports in 3 years jumped from 5% to 13% (1988), of which Cocoa Beans claimed the lion share while manufactures contributed only 1% of total proceeds.

This skewed development poses a big challenge to Nigerian engineering technology. This paper discusses strategies for lifting technical export contribution from the present infinitesimal level to 50% of total export by the year 2,000.

Made-in-Nigeria automobiles, computers, indigenous technology products etc. organised engineering consortiums and bureaux - for securing overseas employment and World Bank contracts utilising Nigerian expertise were identified as being of export value. Other sectors, services and goods identified include iron and steel as well as our comparatively cheap labour - for full exploitation for exports as major foreign exchange earners that should be fully exploited.

Although ECOWAS is the immediate export field, the ultimate target is the multi-million dollar market of the developed countries.

WHY EXPORTS?

General

Apart from the fact that absolute self-sufficiency is Utopian, no country is an island unto itself. Out of the quest for international relations arises international trade, the real interpretation of which is

Export and Import. The simple meaning of export and import is to send to and bring goods and services from abroad.

Though export trade arises as a natural process in international relations among nations, like any worthwhile venture, the decision and action of engaging in international trade must be based on justification. For instance, current records show that Nigerian Banks in 1988/89 committed N533 million to export financing. There must be good reason for this huge investment. So why export, and what should be exported?

The reasons or needs for exporting can be classified into three, namely national, company (commercial enterprise), and personal interests.

National Interest

- Socio-Economic Survival and Development
- Balance of Trade
- Foreign Debt Servicing

Business Interest

-   Procurement of raw materials, machinery, spare parts, and services

Personal Interest

Payment of professional dues, school fees, medical expenses, pilgrimage, and vacation.

WHAT TO EXPORT

Current Export Pattern

Nigeria's total exports in 1987 valued at N30.24b, more than tripled the level of N 8.92b recorded in 1986. Of the 1987 exports, crude oil accounted for N28.739b which is 95%. In spite of the 13% fall in crude oil shipment over the period, the phenomenal revenue improvement was achieved due to a steady rise in oil price from $14.85 per barrel in 1986 to $18.92 in 1987, coupled with the depreciation in the value of the naira.

Furthermore, Non-oil export increased to N1.501b in 1987 from a figure of N0.552b in 1986.

The over 100% increase cannot be attributed to the fall in the Naira value alone, but mainly due to the new Non-oil export drive and incentives introduced by Government, such that in 1988, out of $2.096b total export revenue, non-oil exports netted $239m or 12.9% compared to 5% in 1987.

However, for the purpose of engineering and technology, the joy of the rapid growth of non-oil export cannot be said to be full, because the exportation of major agricultural commodities claimed virtually the whole growth figure. Cocoa beans alone claimed 50% of the growth, followed by natural rubber, palm kernel and tin in order of significance. Out of the N1.501m non-oil export revenue in 1987, the share of N105.1m i.e. *7%* only claimed by *engineering technology* and miscellaneous goods/services is uncomfortably low.

This is the great challenge before us all.

Exportable Engineering Services

For the purpose of this paper, the broad range of engineering and technological activities apart from the physical real products of manufacturing or construction are regarded as services.

Skills Export

Nigeria being a populous country has a large pool of experienced Professionals in various disciplines including engineering and technology – with varied training background from all corners of the world. The post-independence and oil boom era opened the floodgates of rapid industrial growth which has put Nigerian Professionals to test.

Happily, these Nigerians have proved their mettle to the extent that Nigerians have ably filled positions of responsibility from the bottom to the top in many engineering and technological concerns. Therefore, the marketability of Nigerian technological skills is not in doubt.

However, after the glow of the oil boom, the persistence of the economic recession over a decade now has created the largest pool of unemployed engineering personnel in Africa – young graduates from Universities and Polytechnics as well as high calibre Professionals who have been retrenched from declining corporations such as the Railways or liquidated industrial concerns such as the Leyland Vehicle Plant that cannot survive under the present Structural Adjustment Programme.

Although a good number of Nigerian professionals over the years have been performing effectively in the United Nations agencies, technological institutions and industries in Europe and the USA, the home economic conditions has prompted the exodus of Professionals, more than ever before recorded, from the shores of Nigeria. This exportation of skills, albeit transient and sporadic has brought immense relief to the ever growing Nigerian labour market and created a wider channel of foreign exchange earnings for Nigerians. Even the Prof. Ibidapo-Obe Brain-drain Investigation Panel appointed by the President received representations and carried out studies on the pros and cons of exporting Nigeria skills and concurred that in the long run it is beneficial to the country and should not be discouraged.

In order to consolidate the current opportunities being offered in Europe and America, and to harness and fully exploit the untapped opportunities in the less privileged countries, it is hereby proposed that an institutional framework should be established for the proper organisation and overseas marketing of Nigerian expertise in engineering and technology.

The name *Professional Overseas Services Bureau*, with sponsorship by private concerns, the Nigerian Society of Engineers and the Chambers of Commerce is suggested.

The key lines of expertise to be marketed are Education, Engineering Design, Project Management, Plant Development, Engineering Management, Small Scale Industry/Intermediate Technology and Tero-technology under any engineering technology discipline.

The proposed Professional Overseas Services Bureau (POSBUREAU) should not be confused with the Federal Government's Overseas Technical Aid Corp (TAC) which establishment is commendable as a right step in the same direction. The two schemes are complementary but distinct.

The TAC is limited to young graduates whereas the proposed POS Bureau is meant for experienced professionals who cannot be gainfully absorbed by the Nigerian economy. The benefits of gainful livelihood for the individual, repatriable foreign exchange and improved skills that could be brought back to Nigeria are too enormous for the proposed POS bureau to be ignored by Government, the professionals especially the Nigerian Society of Engineers and the Chambers of Commerce.

Mechanical and Agro-Allied Products

(a) **Agro-Allied Products:** The prominence of agriculture in the life of the nation cannot be over emphasised. It is no surprise therefore, that there are so many research Institutions involved in the development of agriculture. To name a few, we have NIFOR (for oil palm), CRIN (cocoa, kola, coffee, etc). Forest Research Institute, Root Crop Research Institute, Stored Products Research Institute, Rubber Research Institute, Livestock and Leather Research Institute, Grains Research Institute and the International Institute of Tropical Agriculture. Though the primary interest of these Institutes is to develop high yield, fast harvest and pest resistant variety of food and cash farming or planting and harvesting per se, some of the institutes have gone into fabrication of the necessary implements and machines to realise the objective.

However, in view of the need for mechanised agriculture to satisfy the consumption of Nigeria's teeming population and also export, by far the greatest contributions have come from the Federal Institute of Industrial Research, Oshodi, (FIIRO), the Project Development Agency (PRODA) in Enugu, the Metallurgical Institute, Onitsha and the African Regional Centre for Design and Manufacture, (ARCEDEM), Ibadan.

The performance of Engineers and Technologists in this great Institute deserve commendation. Among other devices we have from all these Institute, ploughs, barrows, planters, harvesters, palm wine tapping device, palm wine pasteurising and manual bottling equipment, ogogoro (local gin) distillation equipment, palm oil extraction machines, garl frying machines.

We also have industrial and domestic starch machine, glucose syrup machine, rice parboiling machine, cribs/barns for maize/yam preservation, corn and other grains flour-milling packaging and weighing machines, soap making machine, sugar (from cane) extraction/distillation machine yam pounder, pressure cooker and other kitchen utensils. Others are soya milk extraction machine, fruit juice extraction machine, sorghum milling machine, solar fish dryers/cookers, sorghum malting machine, kolanut fermentation/sparkling wine machine, yeast making machine, bakery ovens, cocoa products machine, etc.

The development of the agro-processing machines is very strategic for Nigeria's export business. The agro-processing machines, apart from offering the opportunity of small scale industry development, create the much desired ADDED VALUE to the business of agricultural produce export which is often wantonly manipulated by the London Commodity

Exchange to the detriment of Nigeria and other produce exporting countries.

In other words, with the machines our cocoa for instance can be processed into beverages such as Bournvita and Milo which are well preserved and which can compete with similar products on supermarket shelves overseas, unlike cocoa beans which are perishable, vulnerable to price dictation from buyers and which end-products cannot be traced to Nigeria.

With agricultural produce before export Nigeria will earn double the amount of foreign exchange she now earns from the same volume of cash crops export.

All the aforementioned machines are exportable. Though the technological development in Nigeria's immediate export market, the ECOWAS countries, is about the same, Nigeria has a lot of advantages over her neighbours. Assuming that Ghana and Cote d'Ivoire are equally developed in the area of small scale industry implements and products, the following attribute of Nigeria make exporting to those countries viable:

i. Large population, larger home market which supports economies of scale which in turn yields cheaper unit cost of the products

ii. Nigeria's iron and steel industries provide relatively cheaper raw materials for the production of the metallurgical devices, more readily

iii. The relatively higher cost of living in some of the ECOWAS countries makes Nigerian products cheaper or more affordable. (On a trip to Abidjan in 1987, it cost the equivalent of N300 to fill the 50-Litre tank of the Peugeot 505 car with petrol which in Lagos cost less than N20 then).

(b) **Automobile and Allied Products:** The automobile and allied industry is another area of non-oil export that is yet to be well developed in Nigeria.

There are big vehicle assembly plants in Nigeria whose production capacity can cater for the entire West African market and beyond. The major Assembly Plants are PAN (Peugeot cars, mini-buses), VWON (Volkswagen Cars, Vans/Mini-Buses), Bedford/Federated Motors (buses and delivery vans since 1959), Leventis (buses), Mercedes ANAMMCO (Mercedes Benz buses and trucks), Fiat (trucks and tractors) Styer (trucks, tractors and military tanks), Leyland (buses, trucks and Land Rovers) Bongos (mini-buses and vans), and SCOA (pick-up vans).

On the other hand Ghana's Bus Assembly Plants have been operating since the early 1970's, while Cote D'Ivoire produces light commercial vehicles they are complementary but distinct.

OVERSEAS DEVELOPMENT PROJECTS

It is said that charity begins at home. Enterprise, or business as in common parlance, is not a matter of charity. That is why the entrepreneurial businessman sees the world as his own reachable constituency, rather than his home base alone. The real issue is with the whole wide world, which market front provides the best profit? Home or overseas?

This is the crunch of export decision-making especially for the small and medium scale business.

Assuming therefore, that Nigerian Engineers and Technologists are entrepreneurial, what can they do for export as development experts? Going by the low level of development throughout Africa, it is no exaggeration to say that the scope of contribution for these experts is unlimited. Time is overdue for Africans, particularly Nigerians to claim their rightful place in the area of development planning and implementation which hitherto has been *dominated* by Europeans and American experts who have no practical knowledge of the people, the culture and the terrain. No wonder a number of the development programmes schemed for us in London, New York, etc operate, or under-perform year in year out without achieving the desired result. Therefore, with our privileged practical knowledge of African problems, our experts definitely have an edge. This attribute will only be meaningful, provided our experts can apply themselves in the same way as the foreign development specialists.

The tasks involved are, the identification of profitable projects, organising development consortiums of Engineers of various disciplines and other development experts especially finance companies. Development programmes from conception, design, through finance to project management and operations management should be packaged henceforth by the actors themselves Engineers and Technologists. There should be a radical departure from the days when a flamboyant young finance graduate will gather Engineers, et, al, make use of them, and make all the profit over their heads. Engineers should forthwith take venture initiative, organise other experts in consortiums, get the job done and earn the profits.

The point here is that time is ripe for technical activities abroad that will utilise Nigerian Technology and engineering expertise and finance, and in return yield manifold the much needed foreign exchange for our great country.

Due recognition ought to be given here to the leading role of the African Development Bank in providing finance for Africa's development and for masterminding the creation and promotion of the Association of African Consultants. It is pleasant to know that the Association of Consulting Engineers of Nigeria participates actively in the Continental Consultants' body. Also a few Nigerian firms are today engaged in EEC (European Economic Community)-ACP (African Caribbean Pacific) projects.

The entry of ECOBANK, a transnational private concern with high offshore banking capability in ECOWAS States, is also a blessing in this regards. The Bank's offices in Nigeria, Togo, Benin Republic, Ghana and Cote D'Ivoire will facilitate ECOWAS trade without the usual third party problem of correspondent banks in Europe or the U.S.A.

Lastly, the emphasis here is that in the deliberate plan to export our professional services, we should be more business savvy by taking the initiative to explore and exploit development potentials in other African states to the glory of Nigerian engineering and technology.

Exportable Products of Engineering and Technology

For the purpose of non-oil exports, apart from petroleum, everything else would qualify for mention. For the avoidance of contradiction, it should be clarified that non-oil should be taken strictly as non-crude oil, so that mention of plastic based engineering products, such as insulation materials, or blow-moulded containers, plants and liquefied petroleum gas, all by-products of petroleum, may not fall out of context.

The broad classification for discussion is based on the three basic Engineering disciplines, as follows:

Mechanical and Agro-Allied Products
Civil, Building and Related Materials and
Electrical, Electronics and Related Products.

Mechanical and Agro-Allied Products

All the countries in West Africa import virtually all their motor vehicle needs from abroad. Apart from one or two motor assembles in Egypt, Libya, Tunisia and Kenya, the rest of Africa import their vehicles from Europe and Asia. The only exception is the apartheid Republic of South Africa, in which nearly all the Motor Companies of the world have Assembly Plants.

Three years ago, the Volkswagen of Nigeria exported some vehicles to the Congo Republic, and about two years ago, the Nigerian Government donated some Peugeot Pick-up Vans and Steyr Trucks to the people of Burkina Faso. Since then there has been no significant export of automobiles on record. Instead, what we witness is reducing capacity plant utilisation and shut downs in all the plants. The worst hit has been the Leyland Plant in Ibadan which has gone into liquidation and fallen into the hands of receivers.

But unlike our experience here in Nigeria and other ECOWAS Countries, South African automobile Industry has been growing in leaps and bounds due to active export trade with its neighbouring states including the so-called radical ones – Mozambique, Angola, Zimbabwe and Namibia.

The economic pressure of South Africa is so dominant in that region such that even in Namibia under transition, the United Nations had to procure from apartheid South Africa the Land Rovers, Trucks and weapons for its peace-keeping force because of cheaper costs. What an irony of fate for the UN, indeed the whole world to boost the apartheid economy through these procurements. This unsavoury situation should be blamed on the ineffectiveness of the foreign politico-economic policy objectives and programmes of the independent African States, particularly Nigeria with all its proverbial might.

In the last seven years, a combination of import difficulties in Nigeria, the pressure of the death of completely knocked down (CKD) components on the Assembly Plants and the business initiative of some entrepreneurs led to the gradual development of the virile motor spare parts industry which now supplies up to 30% of the CKD components required. By backward integration, a few of the Assembly Plants invest in components manufactured locally. The ever-growing motor maintenance service industry receives some of its supplies from the local spare parts manufacturers. It is pertinent here to call on the Standards Organisation of Nigeria to establish quality standards for automobile parts manufacturing, in order to ensure that consumers get value for money which is the sine qua non for good market and survival of the local parts industry.

Meanwhile, the following spare parts which are exportable on their own, or come as VALUE ADDED on fully built vehicles for export, are being produced in Nigeria: brake pads and lining, steel rims, radiators, truck shock springs, contact sets, gaskets and clutch plates.

Others are windshields, mirrors, seats, paints, plastic parts, dashboards, tyres, tubes, fan belts, batteries, cable harness, and numerous Onitsha, Aba and Nnewl-made spare parts that now threaten and in future, will replace those from Taiwan, Brazil and Europe.

The notable pioneer spare parts companies are: Onwuka-NAGSMI, Onitsha Roadstar Limited (which exports barrow tyres to Japan), Ferodo, Zodiac, Berec, Yamaha, Addis Engineering (for 3-wheel vehicles), Metrol, Ibeto, Nigerian Machine Tools, the Steel Plants and Nigerian Engineering

Works. The Chambers of Commerce, the Export Council and the Federal Ministry of Commerce should promote more of these potential foreign exchange earners and pave the way for aggressive international marketing of Nigerian automobiles and spare parts.

Civil, Building And Related Materials

The low level of development of Africa in general, offers abundant opportunity for Engineers and Technologists to apply their talents and prosper. Specific cases of infrastructure - roads, bridges, water resources, public buildings and housing projects as well as the challenge of exporting from Nigeria the vital materials which are locally produced, will be considered.

As far as road construction is concerned, the exportable engineering material is bitumen which is used in making different types of asphalts. Bitumen - a by-product of crude oil distillation - is obtained from the Kaduna Refinery.

Unipetrol Nigeria Limited and NOLCHEM are two notable Petroleum Marketing Companies in the business of blending bitumen into a variety of cutbacks for different civil engineering applications on roads and water resistant felting. Unipetrol, from its 10,000-tonne capacity Plant in Port Harcourt, has been exporting bituminous products to Europe, the Middle East, the U.S. and Cote D'Ivoire.

From its Lagos Plant, trucks are being dispatched with bitumen to some West African countries. In the near future, Chad, Niger and Burkina Faso will join the list of consumers of Nigerian bitumen when Unipetrol's Maiduguri Plant is reactivated.

The real challenge here for Nigerian engineers and technologists is, in efficient operations, maintenance management and spare parts fabrication such that foreign exchange may be saved in importing expertise, minimising imported plants and spare parts, at the same time ensuring that this export commodity flows out smoothly, for the much-needed foreign exchange to flow in.

Two major exportable construction materials produced in Nigeria are cement and steel rods. The raw materials and spare parts problems of the dozen

Cement Factories in the Country due to lack of foreign exchange has led to all of them producing below 50% capacity.

In this circumstance, export is out of the question. However, the challenge to Nigerian engineering and technology is the development of local substitutes for the raw materials and spare parts that regularly ground the cement factories. For instance, the use of local refractory materials and skills at the Ewekoro Cement Factory to resuscitate the kilns five years ago should have been commercialised by now. The local refractory technology and other skills involved in cement works, designs, operations and management, apart from saving Nigeria the loss of foreign exchange, can earn foreign exchange for the country if aggressively marketed throughout Africa.

The success of bridge, housing and public building and drainage projects rests on steel rod reinforcements. Leading in the area of export business among the four Steel Mills and rolling Plants is the Delta Steel Plant, Aladja, Bendel State (Edo State). Though this dynamic company has been exporting intermediate steel products to Europe and

Latin America for about 5 years now, not much has been done on the export of steel rods, bars and beams, particularly to ECOWAS States, none of which has the necessary facilities to meet its needs.

In order to provide water for all by the year 2000, the World Bank and UN Agencies are investing millions of dollars on water projects in developing countries which require boreholes, hand pumps in rural areas, filter tanks, concrete and UPVC pipe networks.

The challenge in these development projects for exportable Nigerian engineering and technology is in the following: fabrication of steel casing for boreholes, production of reinforcement for concrete pipes, extrusion of UPVC pipes, foundry casting of hand pumps, and fabrication of filter tanks filled with local sand aggregates as well as the fabrication of simple drilling rigs.

Direct involvement of the NSE incorporated, NBRRI (Nigerian Building and Road Research), consortiums of Nigerian Engineering Consultants and contractors in securing development contracts for housing, water, roads, and drainage systems from the World Bank, the UN, the International Bank for Reconstruction and Development and the Nairobi-based Shelter Afrique will fetch the country millions of much-needed dollars in exchange for Nigerian products, skills and services.

Electrical, Electronic And Related Products

The broad categories of electrical, electronic and related products and services that now occupy positions of indispensability in our national and individual lives, are heavy current or light current engineering. Although there is a wide range of products, materials and services too many to list here, it is remarkable that a substantial variety of basic popular and essential goods and services are produced in Nigeria.

Among the valuable exportable products/ services of heavy current engineering, and electricity (being exported to Niger Republic for over a decade), generators, transformers, high tension and low tension switchboards, high and low voltage cables, wiring cables, concrete and steel poles, switch fuses, electric meters, luminaires, alarm panels, refrigerators, air conditioners, cookers, pressing irons, conduit pipes, voltage stabilizers, sockets and plugs.

The exportable light current products produced in Nigeria include, television sets, radio sets, video players, loudspeakers, alarms, telephone sets, small electronic PABXs, radio-telephones, microwave ovens and personal computers.

Though the new materials or components used in the production of the electrical goods are imported, there is substantial local VALUE ADDED in many cases apart from employment generation. In certain cases, the entire investment and management of the electric/electronic companies are Nigerian, e.g. Adebowale Electrical Industries, Maiden Electronics, JOAS Electronics, MUNI-ENG, ADSWITCH, and Modular Computers. The exportation of these goods therefore, will fetch net benefits in foreign exchange in return for Nigerian skills in engineering and technology.

However, there remains in this realm a big challenge for the Raw Materials Research and Development Council, the Universities and entrepreneurial Nigerian organisations, in the onerous task of increasing local material content to a significant percentage in Nigeria's electrical goods.

NIGERIA'S INDUSTRIAL AND EXPORT POLICY

Industrial Policy

The full realisation of the non-oil export dream depends on a sound industrial foundation for the country. The industrial policy of Nigeria published in early 1989 spells out the policy objectives, strategies, incentives, guidelines and institutional framework on which the success of the export-oriented industry and enterprises depend. Covered in the documents are objective and strategy of increased export of manufactured goods, objective and strategy of improving technological skills, as well as objective and strategy of increased local content.

Export Promotion Policy

Decree No 18 of 11th July, 1986 introduced the following export promotion incentives

(a) Import Duty Drawback

(b) Export Licence Waiver

(c) Export Credit Guarantee and Insurance Scheme

(d) Export Development Fund

(e) Export Adjustment Fund

(f) Foreign Input Policy

(g) Rediscounting of Short Tem Bills for Export

(h) Capital Allowance

(i) Tax Relief on Interest Income, and

(j) Foreign Currency Domiciliary Account

CONCLUSIONS

1. In order to meet Nigeria's need for foreign exchange in the face of unrealised returns from crude oil export, and unsteady yield from perishable agricultural produce export, the diversification into the exportation of manufactured goods poses a big challenge to Nigerian engineering and technology.

2. Among the wide range of machines, manufactured or processed products from our research institute, especially FIIRO, PRODA, the steel plants, private craft shops and industries, are the following exportable items: Motor vehicles. Tractors, 3-wheel vehicles, garl-making machine, solar cookers, kola fermentation/sparkling wine machine, textiles, travel goods, electrical goods, bitumen, iron rods, tools and spare parts.

3. One of the strategies for exporting Nigerian engineering and technology is to establish professional consortiums, co-operatives, and bureau for the pursuit of overseas jobs be it employment or contract especially on

international projects of housing, roads, drainage and water resources development being sponsored by the ADB, ECOWAS Fund, the UN Agencies and the World Bank.

4. The Chambers of Commerce and the Export Promotion Council should embark on special export education programme for Nigerian manufacturers to take advantage of the numerous export incentives introduced by Government. It should also come to the realisation of all that exportation means catching a share of the market overseas to complement home market and profit with the ultimate result of higher plant and manpower utilisation by each company and the nation at large.

5. Nigeria's long range export thrust should be to penetrate the developed world markets, and should not be limited to markets in ECOWAS and developing countries, virtually all of which are debt-ridden and pursuing SAP (economic structural adjustment programme) schemes of self-reliance.

6. Time has come for Engineers to be more business savvy by acquiring finance skills, and by packaging finance with their project proposals in order to compete favourably in marketing their expertise abroad.

7. The future of Nigeria's manufactured goods exports lies in the hands of Nigerian Engineers, as well as indigenous entrepreneurs and industrials such as Doyin Electronics, JOAS Electronics, Bongos Automobiles, Adebowale Electricals, Onwuka-NAGSMI Auto parts, Metro Batteries, Ibeto Groups, Ugochukwu Group, Roadstar Technical Industries and Modelor Computers.

8. The expectation by the year 2000 is that Nigeria's technical export's present share of 7% of non-oil export would be above 50%. At that time the pride of the achievement or the shame of the failure will be yours, mine and all Nigerians.

Chapter Five

IT'S NOW OR NEVER
– CONNECTING DIASPORA AND HOME
Professionals For 21st Century Development

A COMPELLING NIGERIAN – CANADIAN
NARRATIVE ON EMIGRATION,
EDUCATION, ENGINEERING, HEALTH,
CAREERS, PROFESSIONAL AND BUSINESS
OPPORTUNITIES

OBJECTIVES

- To share with colleagues and friends in the NSE family and members of the public personal experiences as one among many engineers in Diaspora - professional and socio-cultural

- to share the dream, bliss and paradox of emigrating from home, including culture shock for some and better life for others
- to draw lessons of ethical environmental management in North America versus environmental damage and chaos in the Niger-Delta
- from personal experience to shed light on Cancer Health for men and for their brave spouses.
- to spotlight the value of the Project Management Professional on the international market
- to highlight the highs and lows of the Canadian and Nigerian - educational, professional and business sectors
- To gear up Canadian – Nigerian business linkage and professional networks

HIGHLIGHTS

- Nigeria & Canada – Do You Know?
- Some Statistics
- Environment
- Comparative Analysis of Professional Associations

- Nigerian Emigration – Canadian Immigration
- Giving Back
- Technology Sharing
- Project Management As A Profession
- Health
- Cancer Health – Prostate
- Bridging Opportunities
- The Hell Joke

DO YOU KNOW?

Table 1: Political Economy

	NIGERIA	CANADA
AREA	1 million sq.km	10 million sq.km
POPULATION	130 million	33 million
GEO-POLITICAL UNIT	36 STATES	11 PROVINCES (8 Provinces & 3 Territories)
LANGUAGES	OFFICIAL - English OTHERS - Many	OFFICIAL - English & French OTHERS - Many
CHARTER OF RIGHTS AND FREEDOMS	More Principle than Practice?	Guarantees: Fundamental rights and freedoms of conscience, and religion, of the press, etc.
VOICE OF DISSENT	Ogoni, Ijaw, OPC	Aboriginals, First Nations, Quebec separatists
IMPACT OF DISSENT	Production loss: 20% 400,000bbL @ $50	About 6 more years of pipeline project delay
2011 GDP (1)	$244billion	$1.8 trillion
2011 PER CAPITA INCOME (1)	$1,486	$51,500
ENGINEERS 2011 estimates	40,000	500,000

(1) Source: World Bank Development Indicators

Table 2: Minerals and the Environment

	NIGERIA	CANADA
MINERALS & MINING	Natural Gas, Crude Oil, Tin, Gold, Limestone, Bitumen	Natural Gas, Crude Oil, Copper, Uranium, Gold, Limestone, Bitumen (Tarsands)
KYOTO PROTOCOL	Signatory	Signatory, with a question
O&G ENVIRONMENTAL PROTECTION	Associated gas flaring 80% (Estimated) Pipeline spillage: old age and vandalism	Associated gas flaring 15 % (Estimated) Pipeline spillage: old age

*Canada endorsed the 2009 Copenhagen (non-binding) Accord but has since 2011 pulled out of the Kyoto (emission control)Protocol to escape penalties for failing to meet targets

CHARTER OF RIGHTS

- Canada Guarantees: Fundamental rights and freedoms of conscience and religion, thought, belief, opinion and expression, freedom of the press, and association
- Canada's guarantees: Fundamental Rights - democratic rights, mobility rights, legal rights, equality rights (equal protection and benefits of the law without based on race, national or ethnic of origin, colour, religion, sex, (sexuality), age or mental or physical disability

PROFESSIONAL BODIES COMPARISON

Table 3: Engineering Profession Comparison

	NIG.SOC. OF ENGNEERS	ALBERTA ENGINEERS
Established	1958	1920
Membership	42,000 Engineers (in population of 130 million)	55,000 Engineers (in population of 3.5million.)
Protected Title	Engineer (Engr.) Engineering Companies membership not mandatory	Engineer (P. Eng.) Engineering Companies permit to practice
Continuous Professional Development	Encouraged/ Voluntary	Mandatory/Enforced
Law & Ethics	Not so tested	Tested, Strong
Inclusivity	Yes	Yes
NSE - NIGERIAN SOCIETY OF ENGINEERS APEGA – PROFESSIONAL ASSOCIATION OF ENGINEERS GEOSCIENTISTS ALBERTA		

Professional membership and population figures referred to 2008/2009

DEVELOPMENT OPPORTUNITY ANALYSIS

Table 4: Nigeria and Canada Comparison - 2009

	NIGERIA	CANADA
Universities	80+	66+
Doctors	40,000	71,000
Defense	200,000 Regulars	50,000 Regulars 50,000 Reserve
Refineries	4	30
Flared Gas	80%	15%

EMIGRATION – ROOT CAUSES

- Huge JAMB (Nigerian University Matriculation Admission) Dropouts, High Unemployment, Professional Unfulfilment, Crises
- Canada Needs People – Negative Or Zero Population Growth Rate, Demography, Economic Growth
- The Tedious Process
- Point System, Interview, Accra
- $20,000 Deposit Or Net Worth
- Do Not Sell Your Property
- The Long Wait – 3 Years Or Oblivion
- Landed Immigrant Status
- Immigrant Welcome Centre
- Diversity & Multiculturism
- Secularism Credentials!!!
- Professional Accreditation Hurdle
- Odd Jobs, Cabbie & Co.
- Culture Shock – Isolation, Mind Your Biz, Children (Protection) Services, Neighbourhood Watch, Accent, Grammar - Licence /License, Advise/ Advice. Tire/ Tyre, Mom/Mum, Apparent / Obvious, Roommate / Flat Mate

- Canadian Experience – Warm and Distant, Prejudice, Pretension, Overqualified
- Information & Education – Online Services, Forms, Point Scoring System
- Charter Of Rights
- Citizenship – Right To Vote And Be Voted

PROFESSIONALS IN DIASPORA

- Paid The Price
- Engineers
- Doctors
- Oil & Gas Professionals
- Other Experts
- Price Paid, Reward Earned: Expertise
- Lecturing & Research
- Professional Engineering Practice
- Project Management
- Construction Management
- Business Consultants
- Underlining Strength– Law & Ethics

SOME DIASPORA BEACONS & POSSIBILITIES

- PLAN-ENG CONSULTING INC, AB, NT, NU, Nigeria www.plan-eng.com
- Agbisco Associates, Alberta
- Patrodaniels Project Managers, AB
- TLR Consulting, Canada, Nigeria
- Khanatek Technologies Ltd. AB
- Igla-Desic Systems, AB, SK, BC
- Oladipo Olasoju Chambers, London
- Emmanuel Alade Law Firm, Nigeria and Alberta
- TFI Money Transfers, Toronto, ON & Nigeria
- Global Energy Development Corporation Inc
- Hundreds more do exist
- Development Grants, Charities, Tech. & Business Partnerships
- Charities
- Professional Affiliations / Reciprocity
- Development Grants
- Project Collaborations Technical Partnerships – IPPs, Solar & Wind Gen, Garbage To-Power Packages, Skid-Mounted Refineries, Gas Compression Plants

IMAGE CHALLENGE ABROAD

- Fraud & Scam (419; the book)
- Oluwoleism (forged documents)
- Dictatorships
- Religious Political, Environmental Rights Riots
- Drug Courier
- Kidnapping & Terrorism
- Armed Robbery
- Corruption
  - Ports
  - Security Check Points
  - Regulatory Agencies
  - Revenue Collecting Agencies
  - Bids, Contracts and Procurements
  - Misappropriated Contracts
  - Valuation Frauds
  - Payment Kickback

Engineers must contribute to fight these stigmas –
one person, one step, at a time. Yes We Can

A RAY OF HOPE to hold, to celebrate

- Raji Fashola – Lagos State Governor
- Wole Soyinka – Nobel Laureate
- Chinua Achebe – Intercontinental Literary Fame
- Gani Fawehinmi – People's Lawyer
- Bala Usman
- Major Umar
- Murtala Mohammed –benevolent military leader
- EFCC (Economic Fraud Crimes Commission)
- Madam Due Process
- Madam Ransome-Kuti –women voting rights
- LKJ – Baba Kekere
- AWOOOO
- Barack Obama, the 44th US President – African-American

VISIBLE CHANGES – GROOMING A NEW LAGOS

- Clean Streets
- Street Lights that overcome NEPA (power cuts)
- BRT and Bus Lane
- Joining the Queue at Bus Stops, Using Pedestrian Bridges
- Working Traffic Lights

- Traffic Flow – Oshodi, Mushin, Oju-elegba, Yaba
- Security Cameras
- Reduced cases of robbery
- Collateral impacts – Airport is orderly and virtually Bribe-free MMA and very few checkpoints
- Increasing Citizens' willingness to pay taxes

MOTION OF COMMENDATION

- On the effective, people's administration of
- Babatunde Raji Fashola, SAN – the Governor of Lagos State
- With a Call for Policy and Programme Sustenance

BRIDGING BACK, GIVING BACK

- One who doesn't Care, doesn't Count
- NSE Members – Care, Carve a Niche
- Be identified for a cause
- Be known for something in Society
- Be Champion of Something
- Uplift the NSE Brand (Nigerian Society of Engineers)

- Mentoring the future generation of excellence
- Community Service
  - Charity
  - United Way
  - Big Brothers Big Sisters
  - Food Bank
  - Food Rescue
  - Christmas Child
  - Community Engineers (ref. Engr.V. Maduka)
  - Council for Alternative Policy
  - Transparency International
  - Job Bank
  - Disaster Relief
  - Cancer / Kidney / Heart Foundation - Screening
  - Book Exchange
  - Engineers Without Borders
  - Food Bank / Food Exchange
  - Consumer Advocate / Ombudsman
  - Environment / Development Advocate / Watchdog / Ombudsman
  - NSE (High School Soccer Tournament) CUP
  - TWINNING (Branches -Local/International:
  - Lead Sponsor

- National or Regional, Public Policy Debate on:
- Global Warming,
- Nuclear Energy – Power, Medicine
- Free vs. Affordable Education
- IT / Space Technology Village

TECHNOLOGY SHARING

- Advances in Science, Technology and Engineering
- Power Systems and Privatization
- Mining – Tin, Gold, Clean Coal, Bitumen
- Research / Tech. Incubation to Manufacturing (Universities Technologies Inc.)

COLD REGION ENGINEERING

- Ambient Condition, Materials, Extreme Temps, Snow Load,
- Permafrost, Wooden Structure, Tornados, Earthquake

THE INTERNET

- Google Search, Yahoo Mail, Wikipedia, Open Office, Network, Collaboration,

GoToMeeting, Blogging, Websites, Facebook, YouTube, Tweeters, Google Drive
- Telephone Banking, Internet Banking
- E-Commerce

THE PROFESSION OF PROJECT MANAGEMENT

- Principles
- Processes
- Procedures
- Prerequisites & Standards
- Code of Ethics
- Curriculum
- Audits & ISO Standard
- Project Initiation & Project Charter
- Scope Definition and Management
- Work Breakdown Structure
- Scheduling
- Communication Management
- Risk Analysis and Management
- Cost Management
- Quality Management
- Change Management
- Earned Value Measurement (Cost Control Index, Schedule Control Index)

- Project Closing
  - Product Verification
  - Project Audit
  - Financial Closure
  - Administrative Closure
- www.pmi.org – Project Management Institute offers:
- Development, research, standardization, accreditation, promotion and protection of the Profession of Project Management
- Global recognition, practice, remuneration

HEALTH IS WEALTH

Negligence and poor health can cut short the dream to function, network and accomplish modest or great achievements that are possible through collaboration between professionals and others at home and in the Diaspora.

Keep Fit
- Eat Well Live Well
- Keep In Shape, Remain Beautiful
- Increased Energy, Expend Energy To Create Energy
- Stress Reduction
- Fat –Avoid It, (Chicken Skin), Lose It
- Strengthen Your Heart, It's A Muscle
- Strengthen Your Core, Back, Abdomen
- Strengthen Your Bones

Regular Health Check
- Screening, Screening, Screening
- Body Mass Index
- Cholesterol (Stroke, Heart Attack)
- Eye screening (Macular degeneration, Cataract

- Diabetes (Glaucoma Blindness, Early Death
- Breast & Cervical Cancer
- Colon Cancer
- Prostatitis (Prostate Enlargement)
- Prostate Cancer (What is your PSA number?)

Prostate Cancer Health Awareness
- Risk Factors – Heredity, Geographic Location, Age, Environment, Vitamin 'D' Deficiency, Diet
- PSA (Pro-Sate Antigen) Number, DRE (Digital Rectal Exam), Biopsy
- Treatment Options – Radical Surgery, External Beam Radiation, Radiation Implant
- Side Effects – Incontinence, Erectile Dysfunction, etc.(from personal experience, radiation implant is the least invasive, and presents very low side effects, if any)
- Common public health advice – obtain PSA number, among other checks, biennially from age 40, annually from age 50. There a number of charities that offer *free* cancer screening in Nigeria, e.g. Tunde and Friends Foundation.

SUMMARY

- Limited Opportunities / Limiting Environment - Emigration, Brain Drain
- Nigeria's Credibility Gap / Image – Environmental Abuse, International Fraud, Kidnapping, Security Breakdown, Corruption – Limited Foreign Investment / Development
- Diaspora – Home Professionals Strategic Connection – Individual, Corporate
- Bridging Opportunities – Careers, Professions, Business & Trade
- Self-Reliance, Enhanced Professional Standard, Enforcement, Home Respect, International Credibility
- Yes We Can, Yes We Can attitude inspires possibilities (US President Barack Obama)
- Lagos Governor R.B Fashola shining example
- Hope, Faith, Confidence, Purpose Commitment
- One Person, One Event At A Time
- Stand Against Corruption, Serve Your Locality
- Work, Keep Healthy, Relax, Work To The End
- Who Does Not Care, Does Not Count
- Make Yourself Count Wherever You Are; Home Is Best, offers bigger Impact opportunities.

###

Chapter Six

INTEGRITY – What Matters, What Doesn't

1. MEANING

- What Does Integrity Mean?
- What are the characteristics of Integrity?

According to Oxford Dictionaries Online, Integrity means the quality of being honest and of strong moral principles; the state of being whole and undivided.

The characteristics of integrity can be found in other words, symbols, or phrases that have similar meaning. Mydictionary.com lists the synonyms as Honour, Uprightness, Candour, Goodness, Honesty, Sincerity, Virtue, Incorruptibility, Purity, Rectitude, Abstinence, Modesty, Morality, Temperance, Obedience, Faith, Justice. The common thread or sum total of all these is *doing the right thing*

In various departments of life there are instructions, signposts or rules to guide or direct people to do the right thing.

2. INSTRUCTIONS ON DOING RIGHT

Ten Commandments; all faiths, even non-faith organisations have their tenets and instructions for their followers.
Doing the Will of God
Follow the example of Christ
What Will Jesus Do – reminder bands and bracelets
By-laws, Laws of the land; Criminal Code

3. PROMISE / INTENT TO DO RIGHT

3.1 Promise to God
Baptism
Confirmation
Confession
Communion

3.2 Promise to Self and Folks
Resolutions

3.3 Promise to Spouse

a. Traditional Wedding Vow:

I, (name), take you, (name), to be my [opt: lawfully wedded] (husband/wife), my constant friend, my faithful partner and my love from this day forward. In the presence of God, our family and friends, I offer you my solemn vow to be your faithful partner in sickness and in health, in good times and in bad, and in joy as well as in sorrow. I promise to love you unconditionally, to support you in your goals, to honor and respect you, to laugh with you and cry with you, and to cherish you for as long as we both shall live.

- Wikipedia 02 Nov 2013 / About.com/wedding vows.

People are commonly in attendance as witnesses.

b. **Ecclesiastes 5:4**

'When you vow a vow to God, do not delay paying it, for he has no pleasure in fools. Pay what you vow'.

Due to the omnipresence of God, it is understood generally that He is present even where there are no physical witnesses. This concept and **Eccl.5:4** imply that all vows are made to God or with God as witness. It is a real reminder that vows should not be taken lightly.

3.4 Promise to Affiliates – Employer, Employee,
Agreements
Covenants
Contracts

'If two of you agree about anything you ask, it will be done by my father in heaven'. **Matt 18:19**

3.5 Promise to the State, Office & to We The People
Anthem
Pledge
Oath of Office

3.6 Promise to Professions
a. Ethics
b. Law
c. Hippocratic Oath (doctors')
- 'I will remember that I do not treat a fever chart, a cancerous growth, but a sick human being, whose illness may affect the person's family and *economic stability*. My responsibility includes these related problems, if I am to care adequately for the sick.'

.... **Excerpts from Hippocratic Oath Modern Version: Written in 1964 by Louis Lasagna, Academic Dean of the School of Medicine at Tufts University**

The question is, how many doctors have turned away sick dying people for lack of necessary fees for treatment in today's heavily monetised world?

d. Iron Ring Ceremony (Canadian engineers)
- 'I [your name], in the presence of these my betters and my equals in my Calling, bind myself upon my Honour and Cold Iron, that, to the best of my knowledge and power, I will not henceforth suffer or pass, or be privy to the passing of, Bad Workmanship or Faulty Material in aught that concerns my works before mankind as an Engineer, or in my dealings with my own Soul before my Maker'.

Excerpts of Canadian Engineers' Sworn Obligation; Ref. Wikipedia 02 Nov 2013 / "Information Relevant to the Iron Ring Ceremony ", Compiled by Dr. J. Jeswiet, 22 November 2001

The question is, how many engineers have found themselves in bad spotlight of dereliction of duty – negligence or outright compromise of their obligation as well as the law and ethics of their noble profession?

e. Court Oath of Testimony, Witness
- • Oath:
-'I swear by [substitute Almighty God/Name of God (such as Allah) or the name of the holy scripture] that the evidence I shall give shall be the truth, the whole truth and nothing but the truth'.
- • Affirmation
-'I do solemnly and sincerely and truly declare and affirm that the evidence I shall give shall be the truth, the whole truth and nothing but the truth'.
- • Promise
-'I promise before Almighty God that the evidence which I shall give shall be the truth, the whole truth, and nothing but the truth'.
(Oath is alternatively allowed to be made according to the beliefs of the person concerned)

Wikipedia 02 Nov 2013 / England and Wales "Criminal Procedure Act 1977"

The question is how many individuals – politicians, intellectuals, the poor, the rich, and even legislators –

year in year out are being caught, tried and have to endure the pains and penalty of perjury?

4. CHALLENGES OF INTEGRITY

Money
'For the love of money is a root of all kinds of evils. It is through this craving that some have wandered away from the faith and pierced themselves with many pangs'.
I Timothy 6:10

"As for the rich in this present age, charge them not to be haughty, nor to set their hopes on the uncertainty of riches, but on God, who richly provides us with everything to enjoy".
I Timothy 6:17 (ESV)

'For people will be lovers of self, lovers of money, proud, arrogant, abusive, disobedient to their parents, ungrateful, unholy…. '
II Timothy 3:2

'Remember it is the Lord your God who gave you power to be rich'
Deuteronomy 8:18

'Be wise enough not to wear yourself out trying to get rich. Your money can be gone in a flash, as if it had grown wings and flown away like an eagle.'
Proverb 23:4-5a

'We make a living, by what we make, We make a life by what we give.'
- Winston Churchill

'Keep your life free from love of money, and be content with what you have, for he has said, "I will never leave you nor forsake you.'
Hebrews 13:5

- Pleasures – Orgies, Pornography, etc
- Power – Politics, Authority, Absolute Authority
- Leadership – graft, scandals
- Greed
- Corruption

- Business Deals
- Cutting Corners, Sports, etc
- Tax Evasion / Tax Avoidance - Tax Havens
- Partisanship or patriotism
- Corporate Accountability at home or abroad (Kellogg Bribery scandal, SNC-Lavalin, Quebec Construction scandal, etc.)
- Environment
- Social Conscience / Responsibility (Fair Wage, Fair Price, Standards)

5. DOUBT ON INTEGRITY GREY AREAS?

When there appears to be a grey area, then you fall into doubts. When in doubt, you are being confronted by *scruples* - an uneasy feeling arising from conscience or principle that tends to hinder action;

-(Freedictionary.com) / Wikipedia 02 Nov. 2013; then it is advisable to follow the Lions Code of Ethics.

-'Whenever a doubt arises as to the right or ethics of my position or action towards others, to resolve such doubt against myself'

-Lions International Code of Ethics No.4

When in doubt is when to obtain good counsel.
'Real friends speak the truth to us in love – even if it is not what we want to hear. Jesus always tells us the truth'.

-Our Daily Bread March 3, 2013.

If we don't listen to ourselves, our conscience, the inner voice, etc, can we listen to the good counsel or the truth.

'Truth is bitter' – an old adage

   6. OVERCOMERS TOOLS

Believers, and even free thinkers have available to them some tools to overcome the challenges of integrity.

Don't follow the thoughts of man
'For from within, out of the heart of man, come evil thoughts, sexual immorality, theft, murder, adultery. '
Mark 7:21

Dare to be different
Don't be quick to commit
Be in good company.
-'Birds of the same feathers, flock together' – an
adage
Advocate – Stand up to be counted

Stop!!!
– You are attracted to what you look at, what you
think
Determination
Courage
Pray!!!
Rev. 2:25-29: 'But until I come, you must hold
firmly to what you have. To those who win the
victory, who continue to the end to do what I want,
I will give the same authority that I received from
the Father, authority over the nations. If you have
ears, then listen to what the Spirit says to (the
people) the church'.

In spite of our best efforts in the struggle to be
winners in integrity, is it not intriguing to be
confronted with the reality or truth in **Psalm 16:2** –
'You are my Lord, my goodness is nothing apart
from you'.

On one level, it could be misinterpreted and perplexing, that (we are powerless and) there is nothing we can really do to have or maintain integrity. No, that's not it. The right level to view this passage is from the point of humility; for it is only with God (and the Word) that all things are possible, integrity inclusive

John 14:6 I am the way, the truth, and the life no one comes to the Father except through Me.

Matt 5:6 – True Happiness - on the Mount – Happy are those whose greatest desire is to do what God requires

II Kings 18 King Hezekiah – following the example of his ancestor King David - he did what was pleasing to the Lord; and he won victory upon victory.

Job 23:8-10 – I have searched in the East, West, North and South, but still have not seen God. Yet God knows every step I take if he tests me, he will find me pure. I follow faithfully the road he chooses; and never wonder to either side.

Brothers and sisters, the choice is yours to follow Him or go the other way.

7. SAINT AUGUSTINE

St. Augustine was an early Christian theologian whose writings were very influential in the development of Western Christianity and Western philosophy. He was bishop of Hippo Regius present-day Annaba, Algeria

-Wikipedia the Free Encyclopedia 24November 2013.

It was St. Augustine who raised the question, 'How come evil in a world created by the all-good God'.

In answering this question, Augustine thought that' everything in the universe is good, including that which appears to be evil. Shadow, dark spots, are necessary to the beauty of a painting. Evil then is actually an absence of good just as darkness is the absence of light. The evil in the human heart or the universe is thus there for the purpose of wholeness.

When we are in complete union with God we will escape from the shadows, from the desires of the body; and escape from the consequences. That union with God is to be attained through the love of God as opposed to the love of the world.

###

ACKNOWLEDGEMENTS

CHAPTER ONE

- James Untui, PMP, 'Africa Rising', PMI Network Nov. 2011
- African Development Bank
- Project Management Institute
- World Bank
- IndexMundi.com
- Wikipedia – the Free Encyclopedia
- CBC News/ Walter McKay – Canadian Security Consultant in Mexico
- New International Version of the Holy Bible (Gateway)
- RFI
- Global Trade Information Services Inc – Market Intelligence
- United States Census Bureau
- Addis Ababa Online

ACKNOWLEDGEMENTS

CHAPTER TWO

- American Petroleum Institute Safety Manual 1985.
- Esso West Africa Petroleum Depot Safety Manual.
- Unipetrol Nig. Plc Manual of Depot Installation Operations.

CHAPTER THREE

- Association of Certified Chartered Accountants, UK Certified Diploma Accounting & Finance – Course Study Pack & Programme 1992
- The Portable MBA Robert F Bruner, S. Ventakaraman, and Others, Darden School of Business, University of Virginia, 1993

ACKNOWLEDGEMENTS

CHAPTER FOUR

- The Nigerian Industrial Policy 1986 published by the Federal Ministry of Commerce.
- International Marketing Management by James M. Livingstone.
- Central Bank of Nigeria Annual Report – December 31, 1987
- Central Bank of Nigeria Economic and Financial Review- June 1988.
- Federal Office of Statistics Export/Import Data.
- Management in Nigeria January/February 1989 (NIM)

CHAPTER FIVE
- Citizenship & Immigration Canada
- Nigerian Society of Engineers
- APEGGA
- Bureau of Statistics
- Project Management Institute

ACKNOWLEDGEMENTS

CHAPTER SIX

- Oxford Dictionaries Online
- The Holy Bible New International Version
- Acting Right in Integrity: Living for Kingdom Prosperity by Anuoluwapo Egbedayo
- Mydictionary.com
- Wikipedia 02 Nov 2013 / About.com/wedding vows
- Wikipedia 02 Nov 2013 / England and Wales "Criminal Procedure Act 1977"
- Excerpts of Canadian Engineers' Sworn Obligation; Ref. Wikipedia 02 Nov 2013 / "Information Relevant to the Iron Ring Ceremony ", Compiled by Dr. J. Jeswiet, 22 November 2001
- Lions Clubs International Code of Ethics

EPILOGUE

This collage of essays is a brilliant contribution to the state of the technical and professional development of contemporary Nigeria. Writing without an inclination to a genre, the author is an accomplished engineer with insightful literary skills. The essays are all variations on the theme of efficient strategic approaches and solutions to problems of the Nigerian society in the areas of international cooperation, foreign exchange generation for development funding, business decision making tools and ethics.

Drawing from his wealth of technical and field experience, the author engaged the reader through the erudite mind of an engineer in the form of conceptual framework design approach that enables the analysis of the problems in terms of concepts, proposals, and postulations. The author posits the availability of business decision-making tools and given due ethical considerations, we professionals can accrue benefits from our labour of investments, personally as well as for Nigeria.

To make this happen, this piece of literary and technical work which spanned over 25 years, resulted in rethinking of our approaches to problems. Within this time span, only God knows how many conceptual frameworks, proposals, and postulations have gone down the drains without the benefit of trial and error. If we do not subject a concept, a proposal to a test, how do we know it works?

A thoughtful analysis of the essays drives home their contextual relevance to the professional life in contemporary Nigeria. Have significant changes occurred in our situation as professionals in the last 25 years? Are the problems in those years any different from the present ones? Are we measuring up to the new challenges in strategic and business planning? Are we ready to avail ourselves of this timely opportunity as professionals?

'**Due Season**' is about encouraging meaningful Nigerians from all walks of life to avail themselves of the full range of professional development opportunities that the modern society has to offer. This is going to be an uneasy challenge. The actions and results contained in the pages of '**Due Season**' serve as guiding light for us. We are reminded of the conscious need to reflect on our personal and professional experience and ask for feedback again and again!

The passion of the author for improved performance and work ethics is striking, since it represents the outcome in a society in which a high proportion of professionals do not particularly pay due diligence, care, and attention to their performance. A successful outcome is not beyond the realms of possibility.

The author is a person of pleasing habits and I welcome the opportunity to write this epilogue. We wish him many happy returns of the 60th birthday to be celebrated with his family and friends in good health and prosperity.

-Jonathan Adebola Lambo, MD
Vulcan, Alberta Canada